The Anthropic Principle

Proceedings of the Second Venice Conference on Cosmology and Philosophy

Edited by

F. BERTOLA

Dipartimento di Astronomia, Università di Padova

U. CURI

Istituto di Filosofia, Università di Padova

CAMBRIDGE
UNIVERSITY PRESS

Published by the Press Syndicate of the University of Cambridge
The Pitt Building, Trumpington Street, Cambridge CB2 1RP
40 West 20th Street, New York, NY 10011–4211, USA
10 Stamford Road, Oakleigh, Melbourne 3166, Australia

First published 1993

Printed in Great Britain at the University Press, Cambridge

A catalogue record for this book is available from the British Library

Library of Congress cataloguing in publication data available

ISBN   0 521 38203 3 hardback

The Anthropic Principle states that the Universe has the conditions we observe because we are here. Out of all possible Universes we can only experience the restricted class that this permits observers. The realisation that this is so, has profound implications for cosmology, philosophy and theology, which are explored in this book.

The sixteen contributors joined to discuss and share their theories within the context of science. The result is a unique collection of papers of great value to professional astronomers and philosophers interested in the role of observers in the Universe.

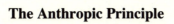

The Anthropic Principle

CONTENTS

PREFACE

The success of the Conference, Kosmos: Cosmology Today Between Science and Philosophy, held in Venice during April 1987, led us to develop the initiative of holding a regular series of encounters between philosophers and cosmologists. The choice of Venice as the permanent seat for these conferences comes not only from the geographical position of this town, an ideal point of contact between different traditions and civilizations, but also from a desire to contribute to the growth of its cultural activities. Venice is known for its history and artistic monuments. It has also been known in the past, and can be again in the future, as a centre in the field of scientific research.

The topics selected for the Second Venice Conference, *The Anthropic Principle,* are among the most interesting and controversial, dealing with both cosmology and philosophy. Without favouring a particular point of view, we invited scholars, experts in disciplines such as Cosmology, Astrophysics, Biology and Theology, and submitted them for an ample discussion. The result was that the Anthropic Principle has been treated in a deep, rich and rigorous way. This is a reflection of the broad extent and significance of current thought in this area.

This volume collects the contributions of the authors.

Francesco Bertola

Umberto Curi

ix

ACKNOWLEDGEMENTS

The following organizations contributed to the success of the Conference:

Regione Veneto
Comune di Venezia
Istituto Italiano degli Studi Filosofici
Goethe Institut
Istituto Gramsci Veneto

Patterns of Explanation in Cosmology

JOHN D. BARROW

Astronomy Centre, University of Sussex
Brighton BN1 9QH UK

Abstract An overview is given of factors that must be taken into account if observations are to be correctly and completely understood. It is explained why 'Theories of Everything' are necessary but not sufficient to achieve this correlation between theory and observation. The essential place of the Weak Anthropic Principle within this scheme is explained.

1. Introduction

In this introductory talk we shall attempt to present an overview of various factors which must be considered before observations of the world can be understood. This will enable us to place the Weak Anthropic Principle in its proper context. Most of what is said is applicable to science in general although where possible we shall focus upon cosmological questions. A subsidiary goal of this talk is to counter the notion that a 'theory of everything' (TOE) of the sort expected to emerge, say, from some edition of superstring theories will explain everything. We argue that although such a 'theory of everything' is necessary for a full understanding of the observed universe it is far from sufficient to complete that purpose. In order to make sense of our observations of the Universe we require an appreciation of seven essential ingredients:

- laws of Nature
- initial conditions
- particles, forces and constants of Nature
- broken symmetries
- organising principles
- selection effects
- categories of thought

We shall make some comments on each of these factors before focussing upon the position of the Anthropic Principle in more detail. More detailed discussions can be found in Barrow & Tipler (1986) and Barrow (1988).

1

2. Laws of Nature

The development of the concept of a 'law of Nature' and what it owes to our social and religious history is a fascinating one which has been discussed at length elsewhere (Barrow 1988). We recognise that there was an evolution away from primitive societies primarily impressed by the irregularities of Nature to those who appreciated the regularities and exploited them to advantage. We have also found it fruitful to view laws of Nature as imposed upon matter rather than to suppose that matter contains certain immanent inclinations which dictate how it should behave. For the Greeks the most perfect laws of Nature were static harmonies of things, but modern science has made progress primarily by viewing laws of Nature as laws of change, even though it is always possible to recast these causal laws of change as static invariances, so that the laws of Nature are reduced to a catalogue of all those things that we can do to the Universe without changing it. The search for the above-mentioned theory of everything' is a search for the set of differential equations which describe these causal laws. Our attitude to the relationship between the laws of Nature (L) and the Universe (U) presents us with the following trichotomy. We can regard $U \subset L$ so laws of Nature are larger than the Universe. This assumption appears implicit in the so-called pictures of 'creation out of nothing' (Vilenkin 1983) in which the Universe begins at a finite past time by some quantum tunnelling event. Such a picture assumes the pre-existence of some underlying structure of physical laws and basic concepts like logic. On the other hand we might assume $L \subset U$ and a theory where, for example, there exist islands of rationality within a possibly infinite Universe. Then again, we could consider $L \equiv U$ so that the laws of Nature are coincident with the material Universe. This rather Augustinian view is implicit in the now unfashionable picture of a Universe that begins at finite past time at a space-time singularity (Hawking & Ellis 1973, Barrow & Tipler 1986, Tipler 1986). Two additional positions could be contemplated with regard to the relationship of L and U: we could entertain the idea that there really are no laws of Nature ($L \equiv \emptyset$) – as do the chaotic gauge theorists (Iliopoulos *et al.* 1983, Froggatt & Neilsen 1989) – or even, as do the solipists, that there is no Universe ($U \equiv \emptyset$)!

Although we require laws of change to understand the Universe we must beware of one awkward feature of our existing laws which creates problems for philosophers. It is possible for different formulations of, say, classical mechanics to have the same physical content but totally different metaphysical implications. Thus, for example, we can use Newton's equations of motion to determine the paths that particles will follow in response to impressed forces or we can use the principle of least action. In the former, causal approach using differential equations, the present is determined by the past but in the latter integral formulation the present is determined by the past and the future. The differential equation formulation leads to a causal interpretation of Nature whilst the least action formulation creates a teleological one. (Indeed the least action principle was introduced by Maupertius in order to justify a teleological interpretation of the Universe as the 'best of all possible worlds' – see Barrow & Tipler 1986 for details.) Thus we must be very cautious in drawing metaphysical conclusions from our pictures of the laws of physics. Incidentally, one should not out of prejudice simply discard the teleological perspective since the path-integral formulation of quantum may yet prove the most fruitful. Moreover, the standard definition of a black hole is entirely teleological since it requires the entire future evolution of the space-time to be known in order to define the event horizon (see Hawking & Ellis 1973, Thorne *et al.* 1986).

3. Initial Conditions

One of the reasons why differential equations are so useful in physics is because they so neatly separate our knowledge from our ignorance. Our knowledge can be coded into the algorithmic content of the equation which dictates how things change in space and time whilst our ignorance is parcelled-up into those factors which we call 'initial conditions' and 'physical constants'. In general, we find that we can use the algorithmic content without having some fundamental knowledge of initial conditions or anything more than empirical knowledge of the constants. Two minor points are worth bearing in mind about initial conditions. First, for differential equations of second order we require the value of a quantity and its first derivative as initial conditions yet the notion of the first derivative requires knowledge of the quantity at times infinitesimally greater than the initial time. Second, it often happens that the influence of initial conditions can be so strong that they can create the impression of their being laws. Thus, for example, the Second Law of Thermodynamics may be primarily a manifestation of the probability of certain starting conditions arising in Nature. We see coffee cups breaking rather than pieces of china moving to create coffee cups not because the latter is incompatible with any 'law' of change but simply because the initial conditions required for the latter are so utterly improbable, whereas those for the breaking cup are rather easy to realize.

In cosmology the issue of initial conditions has become of considerable interest recently. One can identify three distinctive religions amongst cosmologists with regard to initial conditions:

(i) Remove them to the infinite past so the Universe has no beginning in time – although the state of the Universe at past temporal infinity is still an 'initial' condition in the mathematical sense.

(ii)   Show the present state of the observable universe to be independent of the initial state. This was a goal of the steady-state theory (Bondi 1961), the chaotic cosmology programme of Misner (1968) and, most recently, of the inflationary universe scenario (Guth 1981).

(iii) Link laws to initial conditions in some new way so as to render the initial conditions inevitable. This is the approach of the quantum cosmological theory of Hartle and Hawking (1983) who assume a particular 'no boundary' condition to avoid the prescription of initial data in one particular sector of the theory. The idea that some measure of gravitational entropy be a minimum at the beginning of the Universe has also been proposed (Penrose 1987) but this idea has yet to be developed quantitatively. It is hard to see how it could place enough constraints upon the structure of the Universe.

4. Particles, Forces and Constants

Even if one were to know laws of change and starting conditions one would still need to have knowledge of the particles and forces to which these laws apply. The expedient separation of the algorithmic content of differential equations from initial conditions and proportionality constants we mentioned above has the converse that knowledge of laws of change will not completely and uniquely identify the classes of particles to which they apply. However, modern particle physics has departed significantly from the traditional dogma that post-Newtonian science does not want to know 'why' certain particles exist, only 'how' they behave. All the canonical and successful grand theories of fundamental physics (e.g. general relativity, quantum electrodynamics (QED), quantum chromodynamics (QCD) and the Weinberg–Salam theory) are varieties of *local gauge* theory. In such a mathematical theory the imposition of particular space-time symmetries requires that certain forces of Nature and particles necessarily exist with special couplings (i.e. types of interaction). Thus, for example, the invariance under accelerations in general relativity requires the gravitational force to exist. Moreover, certain properties are debarred by the symmetry (for example the U(1) symmetry of electromagnetism forbids a photon mass). However, the search for the most basic gauge symmetries of each of the forces of Nature has failed to tell us why the number of species of particles of a particular sort is limited. Even grand unified gauge theories fail to limit all the populations of particles. Such attempts to unify the description of the different interactions are at present somewhat Platonic in that they are not primarily motivated by observations because of the great expense and difficulty involved in making decisive observations of their consequences. The faith in symmetry as the ultimate guiding principle behind the microworld leads to an 'experimental' approach wherein the particle physicist, armed with the appropriate knowledge of group theory, explores the physical consequences of all the symmetries on offer. Often he may persevere with ones which are blatantly at variance with observation in some particular respect because of a belief that a 'cure' for its particular problem will one day be found and a minor modification will subsequently put the afflicted theory right.

Further factors come into play in such investigations. We need to know what the most elementary particles are. Until a few years ago they were assumed to be point particles (quarks and leptons) described by appropriate quantum field theories. However, these theories are stricken by certain mathematical 'diseases' which are miraculously cured if one assumes that the most elementary objects are linear ('strings') rather than point-like entities (see Schwarz in Hawking & Israel, 1987). There are obvious advantages to be achieved by such a view. In the point picture each particle requires the theory to introduce and specify a different point along with its mass and other intrinsic properties. However, a single string possesses far more information. Different masses can be associated with the energies of all the normal modes of vibration of the string. Unfortunately, the nice mathematical properties of strings possessing all the required symmetries which unify matter and radiation ('superstrings') require a space-time dimension of 10 or 26 and thereby necessitate a mysterious confinement of all but three of the spatial dimensions to very small scales of order 10^{-33} cm. Any variation in space-time of the additional space dimensions would show up as a variation in the (assumed) constant values of our fundamental constants of Nature like the gravitation constant G or the electron charge e (Barrow 1987). These constants, when arranged into pure dimensionless numbers like the fine structure constant, $e^2/\hbar c \sim 1/137$, remain the greatest unsolved mysteries of fundamental physics and cosmology. Their particular values determine in a coarse-grained way the basic sizes and masses of all the structures in the Universe up to the scale of stars. There have been many heroic but incorrect attempts to explain the values of these constants from first principles (see for example Eddington 1946 and Barrow 1989 for a survey of such attempts to explain fundamental constants). Superstring theory claims that one day it might in principle be possible to offer predictions of this sort. Other highly speculative ideas about the wormhole structure of quantum cosmological models offer other possible conclusions.

5. Broken Symmetries

The most awkward thing about the structure of the Universe from the point of view of those charged with its study is the marked propensity it possesses for *broken symmetries*. Although we subscribe to a belief in laws of Nature we do not observe laws of Nature (that is, the equations) in practice: we observe their outcomes (that is, the solutions of the equations). However, these solutions need not possess the same symmetries as the underlying laws. This is one reason why we might expect physics and cosmology to be difficult: we observe only the asymmetrical outcomes of symmetrical laws and are faced with the task of reconstructing the hidden symmetries from those outcomes. Thus we can appreciate why knowledge from any 'theory of everything' can only be partial when we seek to understand our observations of the Universe. We need to know what features of the Universe are fundamental and direct consequences of the laws and symmetries of Nature and which are the quasi-random outcomes determined by symmetries breaking in one particular way and which could (and maybe have elsewhere) fallen out differently. Many large-scale features of the Universe, for instance the number of photons per baryon, or the imbalance of matter over antimatter, or the entire large-scale structure in the inflationary universe theory (see below), may owe their particular forms to random symmetry breakings. It is in such situations that the use of the Weak Anthropic Principle (Carter 1974, Barrow & Tipler 1986) becomes essential if observations of the Universe are to be correctly and fully understood.

Another type of symmetry breaking often occurs in physical systems because they possess very sensitive dependence upon their initial conditions. This was first recognized by Maxwell in the last century (see Barrow 1988, p. 273). In practice, any finite uncertainty in the initial state of such a system renders its future state totally uncertain after a very short period of time. These are 'chaotic' systems. (The weather is a problem of this sort. We think we know the equations which predict how the weather changes with time but we do not have precise enough knowledge of what the weather is like *now* at enough places to produce totally reliable forecasts.) Knowledge of the laws of Nature is not terribly useful in such circumstances and approximate knowledge of the laws can be as good as no knowledge at all when faced with chaotic behaviour. Instead of seeking better and better knowledge of the laws under investigation it is more expedient to seek out those statistical properties shared by all possible laws and equations save for some special set with a zero chance of realization.

6. Organizing Principles

So far we have been talking primarily about the situation which confronts the cosmologist or the particle physicist in the study of the largest and the smallest elements of the Universe. However, there exist fundamental problems in the nether world created by the complicated interaction of many elementary or secondary particles as is witnessed by the wonderful and unexpected complexities of solid state and low temperature physics. Here, and in the problems of the life sciences, one is confronted by the bankruptcy of reductionism. Phenomena like 'life' or 'consciousness' are manifestations of the achievement of a particular critical level of complexity. There is no *élan vital*. Their constitution can always be traced to the stuff which particle physicists study. Biology can in this way be reduced to chemistry, and chemistry to physics, but such a reduction does not permit any understanding to be obtained of the complex phenomena that arise when certain thresholds of complexity are crossed.

Some 'laws' of organization or information-processing are necessary in order to understand how things become organized into complex structures. The laws of electromagnetism are necessary but insufficient to explain the workings of a particular piece of computer software. No TOE suffices. Far from thermal equilibrium the Gaussian 'law of large numbers' no longer holds and the spontaneous highly-ordered structures (e.g. flames) can become probable. Recently, interesting new developments have occurred in this field and have stressed the importance of extending the traditional interests of information theory to describe both the organization of the information within a system and not merely its quantity. One way in which this could be done is to describe the complexity of a state by the running-time or entropy production of the shortest possible program needed to compute the state (Lloyd and Pagels 1988, Chaitin 1987). As yet these ideas have not played any important role in the understanding of cosmology and elementary particles but there are possible cosmological implications of algorithmic definitions of randomness because the most natural of current definitions of randomicity are time dependent. What might be formally random at some early time will not be classed as random later in cosmic history when the amount of time available for 'computation' has increased. We may yet discover fundamental laws of information-processing capability and complexity attainment which must be used to complement any TOE.

7. Selection Effects

Science cannot be based upon observation alone because in the absence of any theory we cannot know what we are observing. More particularly, we need to correct deductions from our observations. Certain types of evidence are more readily obtained than others and the hallmark of a good experimentalist is the ability to understand and foresee these biases as completely as possible. In laboratory experiments their repeatability enables certain aspects of the experimental environment to be altered whilst others are left fixed. In this way many possible systematic biases can be eliminated. However, in astronomy we are less fortunate. We can observe the Universe but we cannot experiment with it.

We are faced with a confinitive rather than an infinitive system and we must try to understand the inevitable selection biases purely theoretically. Thus, if we were to commission a survey of all the visible galaxies with a view to determining their relative brightnesses we would have an inbuilt bias towards finding relatively more bright galaxies than faint ones because they are easier to see in the night sky. Bright galaxies would be over-represented relative to faint ones in our survey and any conclusions drawn therefrom would be similarly biased. The Weak Anthropic Principle is most usefully viewed as the recognition that our own existence imposes a selection bias upon what observations could be made. There are necessary conditions for the existence of carbon-based observers like ourselves and other (weaker) necessary conditions for the existence of any *complex* structures (the necessary condition to be an observer?) at all. If we do not recognize that these necessary properties must be met in our observed universe then we will draw erroneous conclusions from the observations that we make – as would any experimentalist who said he did not care about systematic experimental bias (Carter 1983). A classic example is provided by Dirac's treatment of the Large Number coincidences in cosmology (Dirac 1937a, b, Dicke 1957, 1961, Barrow & Tipler 1986, Barrow 1990). Because he did not recognize that observers are not equally likely to be observing the Universe at any cosmic time Dirac did not realize that one should consider the conditional probability (conditioned upon the existence of observers) that we observe Large Number coincidences rather than the unconditioned probability of large dimensionless numbers arising. The Large Number coincidence between the square root of the number of atoms in the observable universe ($\sim 10^{39}$) and the relative strength of the electromagnetic and gravitational forces between two protons ($\sim 10^{39}$) does not require us to give up Newton's or Einstein's theory of gravity by requiring the Newtonian gravitational constant to change with the age of the Universe. The numerical coincidence of the two Large Numbers is a necessary condition for the existence of observers. If it did not exist nor would we. This shows why we should not think of the Weak Anthropic Principle as an alternative cosmological theory. It is not a theory. It is simply a methodological principle, well-established in other areas of science, that selection effects must be understood, which turns out to give striking conclusions when applied in cosmology because one was not brought up to expect that there exist strong connections between the structure of the Universe as a whole and those conditions necessary for the habitation of an insignificant planet by a life-form of less than moderate intelligence.

The most striking necessary condition for the evolution and existence of observers like ourselves appears to be that imposed upon the size of the observable universe. In order to evolve life spontaneously we require the subtleties of carbon chemistry and the DNA biochemistry that our own genetic code runs upon, along with other astrophysically heavy elements like nitrogen, oxygen and phosphorus. These elements are manufactured during the process of stellar evolution wherein the lightest elements of hydrogen and helium inherited from the first three minutes of the Big Bang are cooked by nuclear reactions into carbon and heavier elements. This process takes of order ten billion years, the main-sequence stellar lifetime, which is determined by fundamental constants of Nature. Hence, because it takes at least ten billion years to give rise to the basic building blocks of atomic life and the Universe is *expanding*, it must be found by observers like ourselves to be at least ten billion light years in size. A universe only as big as the Milky Way galaxy with its one hundred billion stars would have existed for only about a month (Wheeler 1973). The universe would have to be at least as big as it is seen to be if we were the only living beings within it. Of course this does not explain why the Universe possesses the necessary properties for the existence of life; nonetheless, as we shall see below, this certainly does not leave the Weak Anthropic Principle devoid of content. In any cosmological theory which contains inevitable random elements or spontaneous symmetry breakings in its early evolution the Weak Anthropic Principle must be taken into account if a correct interpretation of observations and their relevance to the predictions of any ultimate TOE is to be made. If the Weak Anthropic Principle is ignored answers will be given to non-existent questions and even wrong answers given to fundamental ones.

We have just stressed the fact that there exist necessary conditions for the evolution of observers. According to the biologists there could exist no sufficient conditions. This would be equivalent to teleology – that is, observers would have to exist in our universe – yet we can think of lots of ways in which life could have failed to have evolved by the present or indeed even have failed to evolve by the time all stars died. However, there are unusual conundrums in the interpretation of quantum mechanics which may one day lead us to require the existence of 'observers' (whatever may be their minimum specification) in order to give the universe 'meaning'. The idea that there must exist 'observers' (Wheeler 1973), that there exist sufficient conditions for their existence which must be met, is what Carter (1974) calls the Strong Anthropic Principle although it is best renamed the Strong Anthropic *Proposal* to stress its conjectural status (unlike that of the Weak Principle). If we remove the problem of observership in quantum mechanics by adopting a Many-Worlds interpretation as may be mandatory if one is to interpret quantum cosmology without introducing an 'Ultimate Observer', then it is possible to reduce the Strong Principle to the Weak one. If all possible realities exist then we must inevitably find ourselves inhabiting one of that sub-class which satisfies the necessary conditions for life. We know that this sub-class is non-empty and so there must arise cognizable universes containing observers.

There is an unusual non-quantum-mechanical twist to this question of observership which we should at least mention. Let us consider first the process of computer simulation. In principle we could imagine a simulation being run which was large enough and complicated enough to give rise to observers and so we would have to concede that those observers must be said to exist. It could therefore be imagined that the entire Universe is best thought of simply as the software for an even grander program of this sort for which there exists no hardware. Thus it follows that if observers are possible in any mathematical scheme, and we subscribe to the Platonic philosophy of mathematics (i.e. mathematics exists independently of mathematicians thinking about it), then observers exist in all senses of the word. Since mathematical schemes clearly do exist which give rise to observers, observers must exist.

8. Human Categories of Thought

It has been recognized, at least since the time of Kant's critical writings, that there exists a barrier between reality and our perception of it that is imposed by the categories and modes of human thinking. This barrier may in practice have a very negligible distorting effect upon our perception of reality or it might have a very significant one. Although it is traditional to regard this distortion as large and significant, there are interesting biological reasons why it might be expected to be small. The human brain, like the eye and the ear, has evolved into its present form by the process of natural selection which produces a gradual adaption of sense organs so that they are in tune with reality. Only in this way is the survival value of the organism improved. Thus the ear tells us something about the real properties of sound; the eye tells us something about the real properties of light. For it is the real properties of light and sound that have dictated the course of the evolution of the ear and the eye. A similar consideration with regard to the brain might persuade us that we would expect our mental processes to represent rather faithfully the nature of reality in those areas necessary for our successful evolution.

We have found the most expedient category of thought to use in our study of Nature to be mathematics. Whether this is because the Universe *is* mathematical in some deep Platonic sense, or that we are only any good at discovering its mathematical properties, is a contentious philosophical issue (see Barrow 1988 for a discussion). Yet even within the mathematical perspective we can see that there is considerable scope for choice of different explanatory images. In practice we can build all manner of mathematical images of the reality which we see. In order for these images to be useful we require them to have one primary property: that if we observe *A* to cause *B* then the image of *A* must also, in our mathematical description, cause the image of *B*. This property of *reciprocity* is not sufficient to pin down the representation uniquely. Thus, for example, the representations of quantum mechanics by Schrödinger, Heisenberg and Feynman all possess the reciprocity property.

The unreasonable effectiveness of mathematics (Wigner 1960) in describing the world has different interpretations (or none at all) according to whether one adopts a Platonic, formalist or constructivist (operationalist) interpretation of mathematics. It is interesting to note that the choice that is made can alter what one knows about the Universe. Thus, for example, if one adopts the constructivist philosophy, which accepts as true only those results which can be deduced by a finite sequence of constructive steps and hence *reductio ad absurdum* arguments are invalid, then all theorems which prove the existence of things by contradiction or existence proofs which do not explicitly construct their results are classed *undecidable* rather than true. Thus the singularity theorems of Hawking and Penrose (Hawking and Ellis 1973), which prove by contradiction that there must exist at least one past incomplete timelike curve, would not be 'true' for the constructivist. This dovetails with the operationalist perspective of past cosmic time which regards time as defined by the process of its measurement. Since all clocks defined by physical artefacts are destroyed *en route* to a space-time singularity we could look to the space-time curvature itself as a cosmic clock. In cosmological models like Misner's Mixmaster universe (Misner 1969) an infinite number of physically distinct things (space-time oscillations) happen on approach to the singularity and therefore from the operationalist viewpoint (from which the constructivist philosophy of mathematics has mutated) the initial singularity exists in the infinite past (Barrow 1988).

The most straightforward way to view the effectiveness of mathematics in describing the physical world is to consider the concept of computability, introduced by Turing, from a physical point of view (Turing 1936, Deutsch 1985). If a mathematical function is computable then there exists some configuration of Nature (e.g. atoms, radiation) which can stimulate the operation of that function (as for example, roughly, is the exponential function by radioactive decay). Therefore, mathematics is a good description of the physical world because so many of the simple functions of mathematics are computable functions. If simple functions like addition, sines and cosines, or Bessel functions, were not computable functions then mathematics would not be quite so useful to us in practice. We could still know about the properties of the world that these functions reflect, but only in a non-constructive way. This does not explain why mathematics works so well of course. It merely restates the problem in a more transparent form.

9. The Inflationary Universe and the Anthropic Principle

The development of a refinement of the Big Bang theory of the expanding universe, called the Inflationary Universe (Guth 1981, Linde 1987), has made the
use of the Weak Anthropic Principle quite essential to a correct evaluation of
our cosmological observations.

The essence of the inflationary universe theory is that the early universe
experienced a finite period during which the expansion *accelerated* rather than
decelerated (Barrow 1988a). This is only possible if the matter content of the
Universe is dominated during that period by a source with pressure p and density ρ such that the combination $\rho + 3p/c^2$ is *negative* (c is the speed of light).
Such material 'antigravitates'. As a result of a brief period of evolution of this
type, say only lasting from about 10^{-35} to 10^{-33} s, we can understand why the
Universe is now observed to be so close to isotropy and homogeneity, why it is
expanding close to the critical divide separating 'open' and 'closed' universes,
and why it contains particular small deviations from complete uniformity
which subsequently amplified into galaxies and clusters. The source of the
acceleration can generically be associated with some scalar field existing in a
high-energy physics theory appropriate for the description of the very early
universe. A scalar field φ will contribute a density

$$\rho = \frac{1}{2} \dot{\varphi}^2 + V(\varphi)$$

and a pressure

$$p = \frac{1}{2} \dot{\varphi}^2 - V(\varphi)$$

where $V(\varphi) \geq 0$ is the potential energy of the self-interaction of the field with
itself. Typically we might have $V(\varphi) = k\varphi + O(\varphi^4)$ where $k > 0$ is some fundamental coupling constant. Thus $\rho + 3p$ can be negative if the potential energy
term dominates the kinetic term $\frac{1}{2} \varphi^2 << V$) and the scalar field moves slowly
down its potential surface (if k is very small). The amount of inflation that
occurs will be determined by the speed it moves and the detailed form of $V(\varphi)$.

The result of a short period of acceleration (inflation) is that the entire visible Universe today can have expanded from a much smaller region than it
could in a conventional Big Bang universe which always decelerates. Today,
the radius of the *visible* Universe (as opposed to the entire Universe which
might be infinite) is equal to the product of the speed of light (3×10^{10} cm $^{-1}$)
and the time since the expansion began, and is roughly 10^{27} cm. When the
Universe was 10^{-35} s old what forms the visible universe today would have
been contained within a sphere of radius equal to about 1 cm. Yet, at that early
time the size of the visible universe (that is the distance over which light sig-

nals have had time to travel since the start of the expansion) is only 10^{-25} cm. Therefore the present visible universe is made up of a vast number of regions which must have been completely causally disjoint at 10^{-35} s so it is a mystery why they have given rise to a present-day visible universe which is so similar in its physical properties from place to place. However, the short period of accelerated expansion enables a region of 10^{-25} cm at 10^{-35} s to have expanded by a considerably greater extent by the present so that it can actually encompass the entire visible universe today.

Linde (1987) has stressed that we should consider most likely a scenario of *chaotic inflation* in which we imagine a universe (possibly infinite in spatial extent) at 10^{-35} s in which our scalar field is randomly or non-randomly distributed so that in different places the mix of starting conditions for the evolution of the scalar field and the balance between kinetic and potential energy contributions to ρ and p differ. As a result, each tiny region of size $\sim 10^{-25}$ cm will undergo a different amount of inflation and we could imagine the result as a foam of bubbles – some large, some small – covering an entire random ensemble of possibilities. In such a scenario it becomes essential to take Weak Anthropic selection into account (Barrow & Tipler 1986, Barrow 1988). We could only find ourselves within one of the 'bubbles' that expanded for at least ten billion light years in which the building blocks of biochemistry can be produced. Therefore even if it is improbable for inflation to have produced such a large amount of inflation in a particular model we would still not be justified in discarding that theory because we would have to be existing in a large bubble no matter how improbable its occurrence. One cannot correctly judge the relationship between the inflationary universe theory and observation unless one takes into account the selection effect introduced by our own existence. We might also add to this brief resumé of chaotic inflation the possibility that there might exist a continual temporal sequence of inflationary behaviours. This *eternal inflation* scenario is based upon the recognition by Linde that when inflation occurs in many bubbles, as outlined above, it is easy to satisfy within each bubble the condition that many sub-regions within it should all subsequently inflate and so on *ad infinitum*. Moreover, there is no reason why the first round of inflation we have considered should have been the first in reality so the process could have continued from a past temporal infinity.

10. Discussion

In the previous sections we have indicated those ingredients, in addition to a knowledge of the laws of Nature, which are required if cosmological observations (and indeed any observations) are to be correctly and fully related to our own rational thinking about the Universe. Knowledge of any laws of Nature which we may obtain from some 'Theory of Everything' must be complemented by an equal understanding of those initial conditions, forces, particles and constants, symmetry breakings, organizing principles, selection effects and categories of thought which necessarily augment them to determine what we actually see. The Weak Anthropic Principle plays an essential role within the category of selection effects and is an essential methodological ingredient for the correct appraisal of cosmological theories (like inflation) which possess inevitable random elements.

References

Barrow, J.D. & Tipler, F.J., 1986. *The Anthropic Cosmological Principle*, Oxford University Press: Oxford and New York.

Barrow, J.D., 1987. *Phys. Rev.* D **35**, 1805.

Barrow, J.D., 1988. *The World Within the World*, Oxford University Press: Oxford and New York.

Barrow, J.D., 1988a. *Quart. Jl. Roy. astr. Soc.*, **29**, 101.

Barrow, J.D., 1989.

Barrow, J.D., 1990. In *Modern Cosmology in Retrospect*, eds. S. Bergia & B. Bertotti, Cambridge University Press: Cambridge.

Bondi, H., 1961. *Cosmology*, 2nd. edn, Cambridge University Press: Cambridge.

Carter, B., 1974. In *Confrontation of Cosmological Theories with Observation*, ed. M. Longair, Reidel: Dordrecht.

Carter, B., 1983. *Proc. Roy. Soc.* A **370**, 347.

Chaitin, G.J., 1987. *Algorithmic Information Theory*, Cambridge University Press: Cambridge.

Deutsch, D., 1985. *Proc. Roy. Soc.* A **400**, 97.

Dicke, R.H., 1957. *Rev. Mod. Phys.* **29**, 355.

Dicke, R.H., 1961. *Nature* **192**, 440.

Dirac, P.A.M., 1937a. *Nature* **139**, 323.

Dirac, P.A.M., 1937b. *Nature* **139**, 1001.

Eddington, A.S., 1946. *Fundamental Theory,* Cambridge University Press: Cambridge.

Froggatt, C.D. & Nielsen, H.B., 1989. *The Origin of Symmetries*, World Scientific: Singapore.

Guth, A., 1981. *Phys. Rev.* D **23**, 347.

Hartle, J. & Hawking, S.W., 1983. *Phys. Rev.* D **28**, 2906.

Hawking, S.W. & Ellis, G.F.R., 1973. *The Large Scale Structure of Space-time*, Cambridge University Press: Cambridge.

Hawking, S.W. & Israel, W., 1987

Iliopoulos, J., Nanopoulos, D.V. & Tamaros, T.N., 1983. *Phys. Lett.* **94B**, 141.

Israel, W., 1987. In *300 Years of Gravitation*, eds. S.W. Hawking & W. Israel, Cambridge University Press: Cambridge.

Linde, A., 1987. In *300 Years of Gravitation*, eds. S.W. Hawking & W. Israel, Cambridge University Press: Cambridge.

Lloyd, S. & Pagels, H., 1988. *Ann. Phys.* (NY) **188**, 186.

Misner, C.W., 1968. *Astrophys. J.* **151**, 431.

Misner, C.W., 1969. *Phys. Rev.* **186**, 1328.

Penrose, R., 1979. In *General Relativity: an Einstein centenary survey*, eds. S.W. Hawking & W. Israel, Cambridge University Press: Cambridge, p.87.

Thorne, K.S., Price, R.H. & MacDonald, D.A., 1986. *Black Holes: The Membrane Paradigm*, Yale University Press: New Haven.

Tipler, F.J., 1986. *Phys. Rep.* **137**, 231.

Turing, A., 1936. *Proc. London Math. Soc.* (Ser. 2), 42, 230.

Vilenkin, A., 1983. *Phys. Lett.* **117B**, 25.

Wheeler, J.A., 1973.

Wheeler, J.A., 1987. In *The Physicist's Conception of Nature*, ed. J. Mehra, Reidel: Dordrecht, p.202.

Wigner, E., 1960. *Comm. Pure Applied Math.* **13**, 1.

Anthropic Principle and Ancient Science

ODDONE LONGO

Istituto di Filologia Greca
Università di Padova

I must admit that I am no supporter of the anthropic principle. I think that Livio Gratton's stern, if not pitiless judgement recently given on the topic is true (*Cosmologia*, Bologna 1987, p. 619): we cannot allow the anthropic principle any explicative value, and we must therefore consider it a metaphysical, not a physical proposition if it is to be interpreted as a statement affirming that 'the existence of life is an indispensable condition for the existence of the universe', that is that the constants of nature must have the value they have because otherwise life in the universe would be impossible. The person who is speaking to you is neither a physicist nor a metaphysician, but simply a historian of Greek culture and thought who, as such, does not feel inclined to confuse these two levels of argument (although this type of confusion is evident in ancient thought).

But I must confess that after reading what is probably the *summa* of anthropic thought (John Barrow and Frank Tipler, *The Anthropic Cosmological Principle*, Oxford 1986), or at least after reading those parts which did not require any particular knowledge of mathematics or physics, which I do not possess, my certainties were considerably weakened: one must admit that the illustration of the 'anthropic principle' (or rather, of the various formulations, at least four, of the 'anthropic principle') is alluring or certainly such as to raise the doubt that there is at least some part of truth in that proposition.

But I have another task, and if the organizers of this Congress have asked for my co-operation, it is because the historian of ancient thought is expected to tell the public whether what is being discussed today has already been thought and written by the Greeks. This problem is more and more keenly felt by scientists and contemporary scholars, maybe because the discovery of ancient foreshadowings of modern theories may testify to a certain continuity in the history of western thought, thus legitimizing its working through the ages.

A historian of Western thought, or at least one who is willing to start such a circuit (or sometime shortcircuit), who is asked his expert opinion or advice, may often find himself in a quandary. Scientific modern problems are, in their essential aspects, incomparable to ancient ones, and despite the doubtless continuity in the development of scientific thought they are divided by a theoretical and experimental chasm which started to open not so much with Copernicus as with Galileo, and has successively widened beyond all bounds. But despite this chasm of incommensurability, I think that it is always possible to compare modern and ancient scientific theories at least as far as the great theoretical formulations and general models of knowledge are concerned. And cosmology is one of the fields where keeping the dialogue open between the modern and the old, the Greeks and ourselves is most stimulating and easily approached. In the previous *Kosmos* congress I ventured to propose the Epicurean cosmological model in terms of a 'steady-state universe' and the Anaxagorean one as an 'expanding universe'. It is comforting to read in Barrow and Tipler that the Aristotelian cosmos is the first example of a steady-state universe, although I suspect that Bondi, Gold, Hoyle and Narlikar's theory is more comparable (with several divergences) to the Epicurean, atomistic universe.

But let us tackle the real topic of this debate, and the possible help that a scholar of Greek thought might afford. The above-mentioned work by Barrow and Tipler already provides us with an initial clue, as one of the first chapters, entitled 'Design Arguments', is dedicated to an examination of all the historic precedents of the anthropic theory, in particular to ancient thought. Although I am only a tepid supporter (or even a negator) of the anthropic principle, I can add to the evidence gathered by Barrow and Tipler by proposing a far more significant Greek source, I should say the most significant we have, where we find a formulation of the anthropic principle which is not limited to a general level of teleological and anthropocentric postulations but which encompasses a precise and specific scope.

Before illustrating this evidence (which is Socratic) it is necessary to establish some corner-stones. First of all, to talk about the anthropic principle without any other specification, although constantly done, is not legitimate, and this generic diction is often the cause of confusion and misunderstanding. In Barrow and Tipler's *summa* there are, as I hinted, four different definitions of the anthropic principle, that is: 1. a 'weak' anthropic principle (WAP); 2. a 'strong' anthropic principle (SAP); 3. a 'participating' anthropic principle (PAP); 4. a 'final' anthropic principle (FAP). In our case we shall obviously talk only about WAP and SAP, and more in detail (and less reluctantly) about the former than about the latter. Secondarily, in order to talk about an anthropic principle (whether it be strong or weak) it is not sufficient to ascertain generically teleological anthropocentric formulations, such as are easily found in ancient and medieval thought. There is no doubt that the anthropic principle implies axioms of the teleological and anthropocentric type, and may in a certain sense be defined as the scientific formulation of an anthropocentric teleology. However, in my dissertation related to Greek thought I shall not consider as 'anthropic' those theories which are merely teleological and/or anthropocentric, as I maintain that the core of the anthropic principle, even in its weak form, cannot be limited to that dimension or, in other words, that the anthropocentric and teleological premises are the necessary (but not sufficient) condition in order to recognize the operating presence of the 'anthropic principle'.

Consequently I shall abandon Barrow and Tipler, and all those who attribute an anthropic character to one of the most famous ancient theories, spread above all by Stoicism (and the second book of the *De natura deorum* by Cicero was one of the main means of its diffusion), a theory according to which in all the parts and organs of the human body, eyes, hands, tongue etc., we must recognize a precise finalistic (and teleological) and thus providential design, in the sense that the human organism has been created in this way rather than in another, by a divine providence. But this theory is older than Stoicism, and we first find it expressed in Xenophon's *Commentaria*, where it is attributed to Socrates. Basically, according to this theory the creator endowed man with eyes so that he could see, with ears so that he could hear etc., and also endowed man with an upright position and with hands so that he could better dominate the environment. It is, as one can see, teleological anthropology, but I am rather unwilling to consider it a formulation of the 'anthropic principle'.

En passant, we may notice that the theory we have just expounded was strongly objected to by the ancients, as illustrated by a quotation from Lucretius (IV 822 onwards), when translating, here as elsewhere, an Epicurean text. The passage says: 'beware of mistakenly thinking that eyes were created so that we can see, or arms and hands so that we can use them to our advantage,' etc. (simplified version). As a matter of fact, as an alternative to this finalistic, static model, Greek thought had elaborated an evolutionary model of man and of human societies which excluded any 'providential' intervention on the part of a creator, and saw the above-mentioned human faculties as the final product of a primordial, entirely rudimental and primitive state. So, next to a teleological, statis model excluding any process of transformation such as the Socratic–Platonic and the Stoic model, we have in Greece a model which is at the same time anthropological and cosmological, which excludes any teleological dimension, and acquires evolutionary processes at a physical, biological and anthropic level, even if these evolutionary processes appear to be 'orientated' in an anthropocentric sense (man, or better man organized in society, or better the Greek man of the fifth and fourth century BC, mark the utmost completion of this evolutionary process). This evolutionary model makes its way through Ionic thought where it receives its most complete formulation with Anaxagoras, and is perpetuated, as we have seen in Lucretius' passage, in the Epicurean atomistic theory. On a universal scale we have, as we said before, a 'steady-state universe', that is a universe with a uniform and invariable average density, which has always existed and will forever exist; on a local scale we have, at the level of the single *kosmoi*, or galaxies, or, if you prefer, 'bubbles', processes of atomic aggregation and disgregation which imply both a physical and a biological and anthropological evolution. But, as I mentioned before, I do not think that the 'providential' theory (teleological and at the same time anthropocentric) of Stoicism, and before that of Platonism, has the necessary requisites for it to be included in the category of 'anthropic' theories, unless we are willing to be satisfied with very partial and superficial analogies, or we mortify the anthropic principle reducing it to a generic finalistic issue. The same is true if, of a more generic and blurred Platonic-Aristotelian thought we emphasize the Aristotelian theory, and precisely its two salient aspects of cosmology and anthropology. If we consider the Aristotelian universe as a whole, that is from the cosmological point of view, we are no doubt dealing with a physical universe which is built so as to correspond perfectly to some finalistic purposes; it is a universe which is endowed with a rigid teleological structure, which has no beginning or end in time, and knows no evolution, and at the centre of which there is a 'final cause' which is at the same time its 'moving cause'.

Thus, we have a teleological universe but not an anthropocentric one, considering that it is governed by hierarchies of substances which depend on the 'First Mover'. It is a single universe, and this singleness is the base of its finalistic character, whereas mechanistic conceptions such as that of Epicurean atomism exclude the singleness of the world and admit an infinite plurality of worlds. But, we were saying, the Aristotelian teleological cosmos is not, on the big scale, an anthropocentric cosmos (or else it is, but not programmatically, in as much as it is geocentric). A specifically human teleology is met with in Aristotle on a smaller scale, that is in the sublunar world and in particular (in an anthropic sense) in the absolutely hierarchic organization of the biosphere. In a very well known passage from the *Politics* (I 1 ch. 8) we find one of the most drastic formulations of an anthropocentric bio- and zoological model, where the purpose of the entire system of nature seems to be that of serving the needs and utility of man. Vegetables are created by nature for the use and utility of animals, but animals exist because of their utility to man. All that nature produces is produced 'for man', which is like saying that man is nature's final cause. In particular this teleological sequence works at the level of trophic food-chain. The trophic chain, in its succession of 'autotrophic' organisms (or else producers: vegetables) and heterotrophic organisms at various levels (consumers of autotrophic organisms: herbivorous animals), has its finalistic apex in man, who is a 'supreme' consumer, for whom all other living beings, whether vegetable or animal, exist.

However, even in the case of a construction so strongly finalistic and anthropocentric, it would be very unseemly to talk about an 'anthropic principle', unless one wanted to deprive this principle of any specific theoretical content.

Without attempting an appropriate definition of the 'anthropic principle' in its multifarious expressions, which is not our task, we must stress that we intend to talk of the anthropic principle in the guise of an anthropic cosmological principle, in accordance with the title given by Barrow and Tipler to their treatise. This specification already implies cancelling from the inventory of ancient doctrines much of the material which, in the treatise by Barrow and Tipler, is assigned to our theme.

If we were now asked to give some formulation of what we mean by 'anthropic principle' as a reference and criterion to which theories of ancient thought might be related, we would answer as follows: by anthropic (cosmological) principle, we mean a physical doctrine, a theory of nature assuming the existence of precise and certain physical limits which alone allow the development and perpetuation of life, in particular of human life. Such a theory should presume that the universe, in its present organization, was created as it is now, so as to meet the needs of man's existence and to provide the presuppositions of its possibility, not only as part of a superior whole made of planets, the sun, stars etc. A provisionally unsolved problem is whether this finalization to man (the existence of rigorous and restricted parameters between which life is possible) must be understood in a creational sense or if it is understandable in an evolutionistic perspective which is also obviously endowed with precise teleological features.

It is on the basis of this definition that we believe we may maintain that at the moment of its utmost development Greek thought achieved a formulation of 'anthropic principle' which, considering the rudimental gamut of scientific-astronomical knowledge of that time, appears to possess all the requisites. This formulation is to be found in a passage from Xenophon's Commentaria, which are one of the most precious sources of Socratic thought, given that, as everyone knows, Socrates never wrote his doctrines and his teaching was entirely and exclusively oral.

The topic of the Socratic argumentation (*Comm.* IV 3.8–9) is the characteristics of the solar orbit in relation to the possibilities of life on earth: the astronomical system here presupposed is obviously geocentric, with the sun going round the Earth like all the other planets (although with a regular motion which the other planets do not possess). The Socratic argumentation aims at proving that the entire reality in which we live witnesses the existence of a design, or of a divine providence which presides over it. Now, says Socrates, if we observe the laws according to which the sun goes round the Earth during the year and in the succession of the seasons, we notice that, starting from the winter solstice it comes nearer to the Earth thus starting the process of ripening or drying those fruits for which the ripening time has come. Once this has been done, the sun does not come nearer to the Earth so as not to harm us by giving us more heat than is necessary. The same happens, Socrates continues here, when the sun stops at a certain distance beyond which we would freeze and it then again reverses its course, beginning to come nearer to us. In other words, says Socrates, the solar orbit through the vault of heaven remains within the limits most advantageous to us: this is the proof that all this (that is, the alternation of the seasons deriving from the motion of the sun) is 'for man's sake'. For the same reason the passing from summer to winter, from cold to heat, is gradual, because otherwise our body would not be able to bear either sudden heat or sudden cold. And in another passage by the same Xenophon (*Cyr.* VI 2.29) it is stressed, and it is not Socrates speaking, that the gradual change (the Greek word is *parallaxis*) enables nature to bear alterations: 'it is shown to us by the god' (and by god one must here understand the sun) 'who takes us gradually from winter to bear increasing heat and from summer heat to bear the cold of winter'. One must not forget that the myth of Phaethon was the Greek metaphor of a possible *parallaxis* of the sun beyond the limits compatible with the existence of life on the Earth; the inexpert charioteer could not keep within the course of the solar horses with which his father Helios had entrusted him, and the chariot came too near to the Earth, heating it up and burning its surface.

Really, continues Socrates in *Comm.* IV 3.3–5, the entire disposition of the heavenly bodies seems to serve man's needs: the sun gives us contemporaneously the light of the daytime and, with the diurnal alternation, that of the night which we need in order to rest, thus dividing the day into various hours; and the moon, with its alternate phases, organizes our monthly periods of time. Some centuries later, in the wake of Socrates, God of Prusa (3.73 and following) was asking himself 'What else does the sun do but prepare all that man needs, the seasons, light, the growth of vegetables and animals?' The teleological design is a real and true teleological design; and there is no need for other gods, seeing that for the Greeks who remained faithful to the traditional religion Helios was not a physical entity, a 'burning, incandescent stone', as Anaxagoras had said, but a divinity in itself.

At this point, I believe that we may consider the Socratic theory in a specifically 'anthropic' sense. This meets the above-mentioned requisite of 'a theory of nature assuming the existence of precise and certain physical limits, which allow the development and/or perpetuation of life, in particular of human life'. The seeming ingenuity with which Socrates expounds the doctrine of the binding relationship between the solar orbit and human life, does not lessen the speculative and scientific relevance of this doctrine. Certainly, if we compare it to analogous present-day theories related to the presence of quite clear conditions within the solar system, regarding the existence of human life on the Earth, we face two types of discrepancy: the first, which is also the less relevant, concerns the real motion of the Earth compared to the sun (in the ancient interpretation the apparent motion of the sun during the year was interpreted as real). This discrepancy (geocentric perspective versus heliocentric perspective) does not change the terms of the problem, as what is important is the distance between the two heavenly bodies, no matter which is moving and which is still.

The second discrepancy is to be detected in the fact that, in present-day theories, the distance between the sun and the Earth (thus the mechanism of the Earth's orbit) is only one of the parameters used to define the limits within which human life is possible. It is important to notice that this parameter, once the above-mentioned correction has been made, has remained the same from antiquity up to the present day. Among the requisites for the survival of the human species on the planet, we have to number the regularity of supply of nuclear energy, which must not fluctuate, which requires that not only should the sun continue to burn with extraordinary uniformity, but also that the Earth's orbit be almost circular so as to avoid its getting near or going away from the solar surface. In 1802, W. Paley wrote in *Natural Theology*: 'The permanence of our ellipse is a question of life and death to all our sensitive world ...' As we were saying a short while ago, modern 'design arguments' number a whole series of requisites, and Barrow and Tipler talk on page 87 of 'a copresence of a number of coincidental features in the solar system dynamics and upon which the stability of our environment so delicately hinges'. So that, only to mention one example, the mass itself of the heavenly bodies forming the solar system is comprised between precise parameters, both for what concerns the sun and for planets such as Jupiter, which possesses the maximum mass possible for a planet, beyond which the pressure and the central temperature of the planet would increase so much as to begin nuclear combustion, and the planet would become a star ... (but the same should be said of the opposite case, the moon, which possesses the minimum possible mass below which the satellite would not have the characteristics and the shape it has).

We have to keep in mind this important divergence, namely that the ancient (Socratic) theory establishes one condition only (or one parameter) for the possibility of the existence of human life on the Earth, that is the particular shape of the solar orbit, whereas present-day theories find a copresence and convergence of a whole series of preconditions peculiar to the solar system (not to mention the more general constants of nature). However, I believe that we may recognize in the Socratic doctrine the first true formulation of the 'weak anthropic principle' (WAP), comprised within a wider 'providential' context as the limits of the solar orbit are not only the precondition of human life on the Earth, but they are part of the divine design, aiming at creating optimal conditions for man's existence, which appears to be the ultimate scope of creation.

In order to avoid the deceptive concept of the pre-existence of an 'ancient thought' or of a unitary and coherent 'Greek' thought, we must now briefly recall the theory which is antithetical to the Socratic theory and in general to any finalistic conception of man and of nature: ancient atomism, above all in Epicurus' complete and richly organized conception (unfortunately we know very little of Democritus' atomism), is the doctrine which most energetically opposes any teleological conception of the universe. We have already seen how Lucretius warned the reader 'not to mistakenly think that eyes were created so that we can see ...'; but these antifinalistic positions are valid not only for the human organism, but also for the universe in its totality, and for its fundamental components, and the laws ruling their functioning. In ancient atomism we clash with the systematic refusal to consider the possibility of achieving 'true' knowledge through any interpretation of phenomena in finalistic terms, that is as if they were 'a project'. What I have just quoted is not a passage from a history of Greek thought, but from *Chance and Necessity* by Jacques Monod; Monod recognizes in this refusal, that is in what he calls 'the postulate of Nature's objectivity', 'the corner-stone of the scientific method'.

According to WAP, the number of conditions required for the existence of the universe as we observe it, and for the existence of the planets in the solar system, and among them the Earth with its atmosphere, and for the development of life, animals, man and so on, the number of conditions so that all this may happen is *too high* to be the result of an accidental selection, thus of a non-teleologically orientated evolution: which sooner or later takes us from WAP to SAP. (Talking about evolutionary processes one must keep in mind that some of the constants of nature are *not* evolutionary.)

According to Epicurus, in a spatially and temporally infinite universe, infinite (or rather numberless) combinations and selections are produced and organized according to the rule of 'trial and error'. In a universe which is averagely stationary, but is formed by an assortment of 'bubbles', of single 'island-universes' which are all evolving, the limitations imposed by WAP are annulled by the spatial (and numeric) unlimitedness and temporal infinity assumed: the cosmos and planetary forms necessary to the emergence of life, and life itself, may have evolved innumerable times and with ever different rules.

In other terms, and quoting Stephen Hawking, we may say that, totally against anthropic theories, the universe of ancient atomism is a universe with no boundaries in space, no beginning nor end in time, and with nothing to be done by a creator.

The Anthropic Principle: Laws and Environments

G.F.R. ELLIS
S.I.S.S.A. Trieste, Italy

1. Introduction

The anthropic principle is usually discussed in terms of why the Universe is such as to allow the existence of intelligent life. While it may be taken to imply more, I believe that at a minimum it is a useful tool for increasing our understanding of causal links; I will briefly return to this in the last section.

Given our contemporary understanding of physics, we normally regard present conditions as arising from a given set of physical laws, plus a specific set of initial conditions for those laws (the Machian ideal of integrating the laws and initial conditions into a single unity has not yet been attained in a satisfactory manner). The overall situation is thus as indicated in the following diagram.

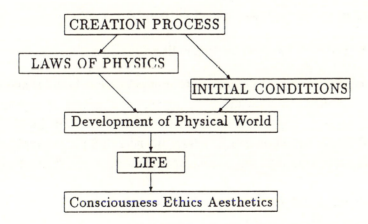

However, the distinction between laws and initial conditions is not always explicitly made in examining the anthropic issue. In my view, one can somewhat refine the basic anthropic question by explicitly posing it in two parts:

(1) Why are the laws of physics such as to allow intelligent life?

these laws determining the kinds of things that can happen in the universe, and

(2) Why are initial conditions in the universe such as to allow intelligent life?

these initial conditions determining, amongst all possible happenings allowed by the physical laws, those that actually occur in the universe. The point is that we can easily imagine other universes with different physical laws on the one hand, and with different initial conditions on the other. Making this distinction helps us to understand which features of the universe are essential for the existence of intelligent life.

Furthermore, the issue can be sharpened further by contrasting whether given laws or initial conditions

(a)   allow the functioning of living systems as we know them,

assuming suitable environments exist; this is concerned with the viability of the individual organisms as systems, and is in fact a hierarchy of conditions as we consider organisms from a single cell to a human being;

(b)   allow the functioning of environments in which life can exist,

assuming it has come into being somehow, i.e. not taking into account the possible development of the environment or life, and assuming individual organisms are viable; this is concerned with the existence of environments allowing the viability of ecosystems at a given time, and again implies a hierarchy of conditions as we consider organisms of differing complexity; and

(c)   allow the evolution of intelligent life,

which is the most stringent requirement, only possible if both (a) and (b) are true. This is concerned with the development of families of organisms in evolving ecosystems, up to organisms that are self-aware. The latter requirements lead to the 'selection effect' nature of the anthropic principle, for (b) and (c) will not be true at all times and places in the universe; thus we can ask

(b´) At what times and places in the universe can life exist?

and

(c´) At what times and places in the universe can life have evolved?

Note here that the distinction (1)–(2) separates causes, whereas the distinction (a)–(c) separates effects. The contention is that in the process of considering alternative universes and their consequences, clarity is increased by specifically and clearly making both distinctions.

2. The Effect of Differing Laws of Physics

As examples, consider first the laws of physics. The *strong force* is required for (a), (b) and (c), for without it nuclei would not exist; similarly *electromagnetism* is required for chemistry as we know it, and so for the functioning of cells and the human body. Thus it is also essential for (a), (b) and (c).

Is *gravity* essential for the functioning of organisms? No; if it were, men could not survive in free fall (an issue debated before space-travel provided a conclusive answer). While gravity is not needed for the functioning of the human body, or for an environment where life can exist, it is needed for the existence of environments in which life can develop, such as the earth, and so for the evolution of life. However an interesting issue here is whether Newtonian gravity will suffice for this purpose, or if general relativity is necessary. It seems likely that Newtonian theory is adequate (as we know that Newtonian cosmology can reproduce the major features of relativistic cosmology,[1] and Newtonian theory is usually used to study galaxy and planetary formation); maybe from this viewpoint a relativistic gravitational theory, such as general relativity, is an unneeded luxury!

Is the *weak force* needed for the functioning of life? Almost certainly not; it is for example difficult to see how it can be needed for the human body to function. However it is probably needed for the development of environments without which life cannot exist, because it plays an essential role in nucleosynthesis. If we take the stand that because of electroweak unification we should not consider the weak force separate from electromagnetism, then the weak force must be a feature of cognisable universes because the electromagnetic force must be in existence there; however it is at least logically possible to consider universes where these forces are not unified, and then the weak force is not required for (a) and (b) but is for (c).

At a more basic level, *quantum theory* is needed for life to exist because without it the atoms out of which our body is built would be unstable; additionally, quantum physics appears to play a fundamental role in some essential biological processes, for example quantum tunnelling is needed to allow haemoglobin to transport oxygen to our body cells (D. Bromley, *Science*, **209**, 1980, p. 116). Similarly *special relativity* is required for the body to function because relativistic corrections result in the 'conversion' of a 4s electron in copper into a 3d one, thus determining its chemical properties and allowing it to function as a vital micro-nutrient, being required for iron metabolism, the functioning of the nervous system, and for normal bone develop-

---

[1]   Element production would be different, but not necessarily so different as to prevent later formation of stars and planets as we know them.

ment (see e.g. Marston *et al.*, *British Journal of Nutrition*, **25**, 1971, pp. 15–30; Wahle and Davies, *British Journal of Nutrition*, **34**, 1976, pp. 105–112). Similarly the chemistry of vanadium is crucially affected by relativistic effects, and this is also a vital micronutrient (*Proc. Nutr. Soc.*, **33**, 1974, 307–313).

We have briefly considered here the necessity of the *existence* of each of these effects. One can of course, in the well established way (see e.g. Carter's original articles, and B. Carr and M. Rees, *Nature*, **278**, 1979, p. 605), also consider alterations in their detailed nature, e.g. by considering the effects of variation of the 'constants of nature' that determine the detailed action of these various forces; the suggestion is that it is again useful splitting up consideration of the effects into (a), (b) and (c). The conclusion is clear: a fine tuning of the laws of physics is needed in order for intelligent life to exist.

In my view the most delicate feature here is that the existing laws of physics allow the higher levels of order to function. We normally take this so much for granted that we do not consider explicitly the extraordinary nature of laws that allow a single cell, much less ourselves, as solutions: a single cell is extraordinarily complex in its functioning, yet we consist of 10^{13} cells functioning as an integrated hierarchical whole, this functioning dependent in detail on bond lengths and angles that allow the DNA molecule to have its spiral structure, the bases to have their complementary nature, enzymes to function, and so on. Not only do such solutions exist, but they are of such flexibility as to enable us to talk of adaptation in a purposive sense (animals have fur on their feet to keep them warm, wings have hollow bones to allow them to be light, etc.). Very small variation in the fine structure constant, for example, will prevent these intricate mechanisms from functioning. The point here is not how life evolved to have such complex functioning, but that the laws of nature allow this functioning in all its incredible variety: that all the different life forms can be solutions of the equations expressing the fundamental physical laws. Higher levels of organisation can be superimposed on complex structures, as John Barrow has pointed out (and see also e.g. *The cosmic blueprint* by Paul Davies), but this can happen only if the complex structures are flexible enough in their operation. This is where amazing fine tuning occurs in the laws that make this possible.[2]

---

[2] Realisation of the complexity of what is accomplished makes it very difficult not to use the word 'miraculous' without taking a stand as to the ontological status of that word.

3. The Effect of Varying Initial Conditions

Some fundamental features of the way physical laws operate, for example the congruence of the different arrows of time, appear to depend on boundary conditions in the universe. Thus the local functioning of entropy and the consequent development of structures in the universe are both dependent on initial conditions at the big bang (cf. G. Ellis and D. Sciama in the *Synge Festschrift*, ed. L. O'Raiffeartaigh; R. Penrose in *General Relativity, an Einstein Centenary Survey*, ed. S. Hawking and W. Israel; and H. Reeves in this volume).

Given such basic features as the arrow of time, the role of initial conditions is primarily in allowing suitable environments in which life can survive and develop. There is a fundamental requirement here that the universe allows 'isolated systems' to exist, for at its early stages of development, life is too fragile to withstand much buffeting by the environment. Thus life cannot exist unless there is a stage in the history of the universe where for example gravitational waves and electromagnetic background radiation are too weak to disrupt living beings or the environments in which they live. We can easily imagine universes where this is never true, for example in which the background radiation temperature never drops below 300 K; then the functioning of chemical thermodynamics as we know it on earth would not be possible (the dark night sky would not be available as a heat sink and so we would have no waste bin for the heat produced in ordinary chemical processes). Thus only restricted initial conditions allow (b). If they do allow (b), then there will in general be only restricted times at which this can be true.

Even more, adaptation to the environment requires a stable environment to adapt to, so conditions must be such as to allow this, i.e. there must be a stability to which adaptation can respond (incidentally, if there were major interference caused say by a random gravitational wave background, we would be unable to establish the laws of physics by laboratory experiments, for we would be unable to have the required environmental stability).

Supposing these basic conditions are fulfilled, the full requirements for the evolution of life are much more delicate, involving in particular features such as the detailed thermal and atmospheric evolution of planets (e.g. how and when ice ages occur), through to the celebrated case of the expansion timescale, which must allow stellar evolution to take place in a suitably old universe, giving well-known limits on the Hubble constant in cognisable universes. Clearly these restrict the time and/or place in the universe where evolution can occur, as well as the family of universe models within which evolution is possible. An interesting challenge is to see if these requirements could be shown to prove that the universe must be highly isotropic about us (for example, in an inflationary universe of the chaotic type envisaged by

Linde, do they imply we must be situated in a homogeneous regime far from the walls where major inhomogeneity occurs?). Hints in this direction (see e.g. *The Anthropic Cosmological Principle* by J. Barrow and F. Tipler) are more suggestive than conclusive.

The overall conclusion is clear: a restricted range of initial conditions in the universe is needed in order for suitable environments for intelligent life to exist, and *a fortiori* for intelligent life to evolve from a situation where there is no life.

4. Conclusion

I understand the anthropic principle's primary role as being to enable us to comprehend causal links we would not otherwise realise existed. It does so in the context of the requirements for life as we know it; it is very difficult indeed to argue regarding the possibility or otherwise of totally different life forms (i.e. life forms not based on carbon/RNA/DNA, as life on earth is). Given this context, it is clear that both special physics and special boundary conditions are needed (a) for life to function; (b) for environments to exist allowing its functioning; and (c) for its evolution from an inanimate world.

It is clear then that we can use our understanding of these causal links to understand selection effects, restricting those universe models which are cognisable, as well as times and places in those models where we can exist. If we were clever enough we could further pursue the Bondi argument (see *Cosmology*, H. Bondi, Cambridge University Press) that because of the influence of cosmology on local physics, by understanding fully a restricted portion of reality we would understand the whole universe; in particular, by understanding fully the implications of intelligent life we could predict virtually all the large scale structure of the observable universe. We are presently a long way from this goal; but we can still consider

How much can we deduce about the universe in the large from our own existence?

As discussed elsewhere in this volume (and see also G. Ellis, *Gen. Rel. Grav.*, **20**, 1988, 497), one can try to take the implication of the anthropic issue much further, towards ontological questions; this will inevitably lead to controversy. However it seems to me that the use of the anthropic line of enquiry in the more restricted sense advocated here is indisputable.

Whichever line one takes, I hope the examples above show that a useful sharpening of our understanding of the causal links is gained by separating the causes and effects into the different categories discussed here.

I thank H.J. Kreuzer for discussions on the effects of the laws of physics on chemistry.

The Anthropic Selection Principle and the Ultra-Darwinian Synthesis

BRANDON CARTER

Department of Relativistic Astrophysics and Cosmology
CNRS – Observatoire de Paris, 92 190 Meudon – France

Abstract The 'anthropic principle' (in its ordinary 'weak' form) is presented within the framework of the Bayesian paradigm as a 'means of sharpening Ockham's razor' when self selection effects are relevant. It is shown how anthropic selection should be considered as an adjunct to ordinary natural selection, explaining features of our evolutionary history that ordinary (neo) Darwinian theory is incapable of accounting for by itself.

1. Introduction

This contribution consists of three parts, the first of which (sections 2 and 3) describes the fundamental strategy of the scientific method, with particular emphasis on the 'Bayesian paradigm' whose explicit or implicit use is widespread throughout science (as a commutative compromise between the old fashioned 'Determinist paradigm' and the non-commutative 'Heisenberg paradigm' of modern quantum theory). The Bayesian paradigm plays an essential role in achieving the main purpose of this contribution, which is the clarification, in the second part (sections 4 and 5), of the scientific status and meaning of the *anthropic self selection principle* in its ordinary ('weak') version.[1] The purpose of the third part (sections 6 to 8) is to illustrate the use of this principle and to demonstrate its importance by describing one of its more recent and, to me, most significant applications.[2]

Early applications[3,4,5] were concerned mainly with primarily astronomical or cosmological questions, for which the anthropic aspect was essentially an accessory though none the less indispensable element in the correct understanding of matters pertaining to the astrophysical field. The more recently developed application that is described here is primarily concerned with the opposite, i.e. biological, point of view, the astrophysical aspects being treated as merely accessory, though still indispensable, to the primary question of the nature of the evolutionary mechanism that led to our own emergence as a scientific civilisation on this planet. Our evolution is analysed here in terms of what may appropriately be described as an ultra-Darwinian framework, meaning a synthesis of the traditional (local) Darwinian concept of natural selection (acting among genetic strains within a single connected biosphere) with the more recently developed (non-local) concept of anthropic self selection (acting in this case among entire disconnected biospheres, in favour of those on which scientific observers ultimately evolve). Whereas the ordinary Darwinian selection process may be encapsulated by the catch phrase 'survival of the fittest', the extended ultra-Darwinian process also involves both an intrinsic and an extrinsic anthropic selection mechanism, the latter, which is the main subject of the present analysis, being analogously encapsulated as 'arrival of the quickest'.

Quite apart from its interest in the context of the search for extraterrestrial life-systems and civilisations,[6] the use of this extended ultra-Darwinian picture would seem to be absolutely necessary for a proper understanding of what has occurred in our own particular case. A narrower Darwinian or neo-Darwinian framework of the conventional kind is hard put to account for the much discussed appearance of systematic purpose in the fossil record of our own evolutionary history, except by explaining it away as an illusion. However in an extended ultra-Darwinian framework such appearances of teleological 'design' can easily be conceived to be explicable on a respectable causal basis as something of genuine significance for our natural history.

Before starting the discussion (in sections 6, 7, 8) of this particularly important application of the anthropic principle to the history of our own life-system on earth, I have provided a self-contained general purpose account (in section 4) of what this principle is supposed to mean (at least to me) and of how and to what extent it can be justified, taking a little more care than on previous occasions to show how it is related to the even more fundamental principles of scientific methodology that were anonymously expressed (by someone regarded by many of his surviving contemporaries as having been the outstanding physicist of their generation) in the following informal, off hand, off the cuff, but not quite off the record, confession:[7] '... *When we do not know something, we have the theory of probability, indefensible as it is, to help us ... we must start out with an* a priori *hypothesis ... all states that take too many words are equally likely, but ... we have given a little extra weight to some special state ... we observe ... and use the Bayes theorem backwards ... we calculate, according to our hypothesis for weighing ...*'

This hesitant impromptu synopsis of the philosophy of science by a notoriously anti-philosophical scientist is what I have tried to restructure in a slightly more formal way in sections 2 and 3. However, it would be difficult, without getting into the technical details of examples, to put it much more clearly than in the extract above. Just two basic 'principles' are discernible therein, only one being considered sufficiently 'indefensible' to need justification by reference to a named authority: the more or less innocent 'fall guy', who is left to take the responsibility for what greater men can no more do without than defend (though Keynes[8] once made a brave attempt), being the obscure eighteenth-century clergyman, Thomas Bayes (whose framing seems traceable to Laplace[9]). The Bayesian induction procedure has been (prior to the introduction of its even more controversial quantum analogue), at least implicitly, the vehicle of virtually all scientific analysis of empirical results, but it needs to be used 'openly'[10] in the rather delicate cases with which we are concerned here, and is recapitulated for this purpose in section 2. This presentation makes use of the concepts of (additive) 'debit' and 'credit', the latter being a technically useful function of ordinary 'probability' (not just a euphemism like the term 'propensity' that is sometimes used in place of 'probability' by those who dare not use this 'indefensible' concept 'openly').

The less controversial and arguably most fundamental of these basic principles (whose applicability is not limited to science in the strict sense but extends even to theology) is the one commonly associated with the name of an even earlier, though not so innocent, cleric called William of Ockham, an eloquent advocate of the common-sense strategy of giving *preference to the most economical hypothesis*. Although this 'razor' principle (discussed in section 3) is easy to formulate 'bluntly' in vague terms, it is not always so easy to interpret in a 'sharp', unambiguous manner when it comes to actual applications. The anthropic principle (in the ordinary 'weak' form with which we are concerned here) can be construed (as explained in section 4) as a natural and generally appropriate prescription for 'sharpening the razor', i.e. limiting the ambiguity in the interpretation of the Ockham principle, in certain circumstances when one is dealing with effects that are correlated with the presence or the very existence of whoever may observe them.

Whereas my original, less thoroughly developed formulation[1] was sufficiently clear to have been quite easily understood and very soon applied by many working cosmologists,[3,4,5,11] this was essentially due to its having been illustrated by concrete examples (including notably that of the implicit use[12] of the anthropic principle before its explicit enunciation); nevertheless my condensed formal statements of the principle (the first of all simply being '*what we can* expect *to observe must be restricted by the conditions necessary for our presence as observers* '[1]) were insufficiently explicit to convey their complete meaning when quoted out of context. As a result it has been common to criticise the anthropic principle as a mere tautology or even to condemn it as being 'unscientific because irrefutable'. Such criticism is justified in cases such as the much cited but inappropriate example of Chamberlain's historically important lower age estimate[6] for the sun, which is indeed in the tautology category as far as its 'anthropic' aspect is concerned. However, the nontriviality (not to mention the non-irrefutability) of the principle when correctly understood is illustrated by the possibility of deliberately violating it, as exemplified by the forceful assertion of what amounts to the right to repudiate the anthropic principle (in favour of purportedly higher principles) that was made in reply to Dicke[12] by such a distinguished thinker as Dirac,[13] who (while no doubt scientifically unorthodox in his persistently 'rationalist' rather than empirical outlook[14]) is nevertheless (*ipso facto*) above suspicion of irrationality.

The discussion in sections 4 and 5 is intended to clear up the misunderstandings and reply to at least some of the criticisms[15] by distinguishing rather more explicitly than in my original treatment between firstly what might be called the 'anthropic tautology' (which is all that is involved in the Chamberlain example), secondly the 'anthropic theorem', thirdly the ordinary (weak) 'anthropic principle' (that was implicitly applied by Dicke and rejected by Dirac), and lastly the 'strong anthropic postulate' (about which I have doubts of my own) that is of course the least compelling. [Other writers referred to in the comprehensive treatise of Barlow and Tipler,[6] have in recent years made further clarification necessary by extending the 'anthropic' nomenclature to concepts such as 'anthropic finality' whose teleological nature is, as John Leslie has emphasised,[16] quite contrary to the empirical, conventionally 'scientific', spirit of the anthropic (*ex post facto* selection) principle as I intended it to be understood. Indeed it may be remarked that in the classic example of the Dirac–Dicke controversy mentioned above,[11,12] in so far as Dicke's empirically scientific argument clearly foreshadowed the ordinary 'weak' anthropic principle, it is equally evident that Dirac's metaphysically motivated opposing argument no less clearly foreshadowed what has since, rather infelicitously been called the 'final anthropic principle'.[6] I suggest that the confusion[17] would be reduced if, as in the title of this article, the former kind of (*ex post facto*) argument would correspondingly be referred to as the 'anthropic *finality* principle', the more extreme variant[6] advocated by Wheeler being called the 'anthropic *participation* principle' or maybe the '*strong* anthropic finality principle'.]

The 'strong' postulate mentioned above was referred to as a 'principle' in my original nomenclature, following the rather loose usage in the example of the 'cosmological principle' which is actually an ordinary scientific hypothesis (postulating approximate large-scale homogeneity and isotrophy), subject to observational confirmation. With the encouragement of John Barrow, and in a further attempt to reduce the confusion resulting from insufficiently precise nomenclature, I have reserved the term 'principle' here for a rather stricter usage, in which one distinguishes the category of optional hypotheses and consequent laws from less specific but more widely applicable 'principles' whose imposition is more or less obligatory at the outset in entire classes of theory. Just as in the organisation of a state there is a hierarchy descending from a 'permanent' (but often vague) constitution through more easily modifiable (and usually more precise) ordinary laws down to ephemeral (detailed) administrative regulations, so in science there is a hierarchy from general 'principles' through specific 'laws' down to the detailed technical assumptions involved in the interpretation of particular experimental tests. In physics, perhaps the best-known example of a 'principle' in this sense is the entropy generation inequality known as the 'second law' of thermodynamics, which, while not absolutely universal, nevertheless transcends entire classes of physical theory. (The equality known as the 'first law' is of a rather lower ranking, context dependent, type, since energy conservation, although independent of many details, is nevertheless dependent on the – only circumstantially valid – postulate of a stationary background.) In biology the classic example of a principle transcending ordinary laws is the Darwin–Wallace idea of natural selection, whose general formulation is quite independent of the genetic code and mutation mechanisms by which it may be implemented.

The status of the (ordinary) anthropic principle is comparable to that of Boltzmann's entropy principle and Darwin's natural selection principle in that all three are ultimately derivable from what experience suggests to be 'reasonable' general *a priori* assumptions about probability distributions. In the case of natural selection the situation is either too vague or else, if detailed, too complicated for the complete chain of derivation to be easily made explicit, but in the more elementary case of the entropy principle the derivation (via the '*H* theorem') was already analysed in mathematical detail by Boltzmann himself. It will be shown (in section 4) that the corresponding derivation for the anthropic principle is even simpler, requiring just a corollary of Bayes' theorem.

Of course the more transparent the derivation, the easier it is to raise doubts about the assumptions on which it rests. Hardly anyone has seriously called into question the operation of the Darwin–Wallace selection mechanism as such (the polemics about Darwinian theory concern whether it is sufficient by itself to explain all it claims to), but there has been serious speculation about conceivable breakdown of the entropy principle (thermodynamic time reversal) in cosmology. The anthropic principle is even easier to repudiate if one really wants to (Dirac has implicitly given the example) but I nevertheless defend it on the basis of Ockham's principle as the simplest and most straightforward principle to adopt in cases where one has no other reliable guide. The entire history of science encourages faith in Ockham's principle as the safest policy (while the fate of Dirac's non-anthropic alternative hypothesis, of varying gravitational coupling, now rather convincingly refuted by experiment, provides yet another warning against wandering astray).

Having illustrated the meaning to be attached to the term 'principle', I would mention that Hubert Reeves (see his accompanying discussion)[18] and others have made the criticism, which I entirely accept, that the adjective 'anthropic' (which is by now too firmly entrenched to be changed) has unduly restrictive connotations, since the principle in question is intrinsically applicable in the same way, not only by our human civilisation, but also by any extraterrestrial (or non-human future-terrestrial) civilisation that may exist. A less ethnocentric and more informative (but correspondingly less catchy) name for it would be 'observer self-selection principle'.

2. The Bayesian Paradigm

A great 'paradigm' (to borrow the concept developed by Kuhn[19]) acquires its *'titre de noblesse'* by displacing its predecessor (whose status as a paradigm may not have been noticed until then). Although not everyone is willing to recognise it 'openly'[10] without the protection of anonymity,[7] it is hard to deny the effectively paramount status in most of science of what in Kuhnian language may be called the 'Bayesian paradigm'. Although, naturally, a focus of criticism, it nevertheless has no worthy challenger except in the field of fundamental physics, where its heir, the 'Heisenberg paradigm', has already taken over. Both the Bayesian parent and the Heisenberg descendant may (as their opponents claim and many users admit) be 'indefensible', but so long as they have no serious rival they do not need to be. This does not mean that their predecessor and common ancestor is by any means dead: the traditional 'Determinist paradigm' that reigned (at least formally) over science until the present century is still alive and well, but since losing its last great champion (Einstein) it is safely besieged in a circumscribed domain that includes the jury box, but that has as its citadel the impregnable fortress of the primary school.

Both the 'Determinist' and the 'Bayesian' paradigms are based on the (provisionally indispensable if ultimately naive) notion that empirical perception can somehow be described by some logical 'statements' called observational 'results' of the '(I feel hot now) *and* (I remember feeling cold)' type, which are postulated to be mutually consistent within a logical framework of the ordinary Boolean type in which the axioms are isomorphic to those of ordinary set theory (with the correspondence *'and'* $\leftrightarrow \cap$, 'or' $\leftrightarrow \cup$). These elementary 'result statements' are considered as the raw material for scientific analysis, whose role is to organise the elementary results in terms of a logical superstructure involving supposedly more fundamental statements called 'hypotheses'.

Of course, as has been remarked for example by Pierre Duhem,[20] the commonly accepted dichotomy between 'results' and 'hypotheses' is itself ultimately an artifice, and as such might conceivably be eliminated in some improved paradigm that has yet to be developed. This dichotomy is already effectively transcended in the traditional, but no longer paramount 'Determinist paradigm' according to which the type of theoretical organisation to be sought should consist just of a single assembly of 'hypotheses' that are fully consistent both with each other and with the 'results'. (Adoption of the Determinist paradigm suggests, but certainly in no way imposes, the acceptance of the deterministic metaphysical idea of the existence and uniqueness of a special such assembly called 'the truth'.) It can be seen that such a theoretical organisation is just a binary 'yes or no' classification of hypotheses, with the use of the 'yes' label restricted by the consistency requirement.

The 'Bayesian paradigm' is slightly more elaborate (just a little too much so to be suitable for putting over in primary school). Instead of restricting oneself to a mutually consistent assembly, one deals simultaneously with a larger assembly of in general mutually inconsistent 'hypotheses'. The specification of a 'complete theory' now requires that, instead of simply being given an all-or-nothing, yes-or-no label, each hypothesis should be given a corresponding numerical weighting, P say, subject to the requirement that the weightings be attributed consistently with the traditional axioms of mathematical 'probability theory' (as taken over from 'relative frequency' theory) whereby P is interpretable as a 'measure' in the technical mathematical sense. In the simplest version of the Bayesian paradigm, the 'results' are still given a deterministic treatment within the probability scheme by attribution of unit weight to the corresponding statements and hence zero weight to the complementary statements.

In so far as the Bayesian paradigm may be criticised for incorporating a dichotomy between 'results' and 'hypotheses', this objection is no less applicable to the more sophisticated (tertiary rather than secondary school level) 'Heisenberg paradigm'. The latter differs from its Determinist and Bayesian ancestors in admitting 'statements' of a new kind (which may have the status of 'results' or 'hypotheses') for which one relaxes the usual axioms of Boolean logic by dropping the commutativity postulate (though retaining associativity) for the '*and*' operation (commonly denoted by $\cap$ or expressed simply as a dot product) in combinations that might even be rejected as inconsistent in ordinary logic, such as for example '(the spin points north) *and* (the spin points east)'. Like the Bayesian paradigm, the Heisenberg paradigm is essentially dependent on the 'indefensible' notion of 'probability' (which is so embarrassing that people resort to all sorts of euphemisms such as 'likelihood' or 'propensity' rather than name it openly), but without the set theory axioms of Boolean logic, the probability weightings can no longer be assigned in the form of an ordinary mathematical measure. Instead of using a no longer tenable analogy with *sets*, the Heisenberg paradigm associates observable 'results' (and other statements) with *projection operators* on a Hilbert space. (Note that the product of two projection operators will also be a projection operator if they commute, but not in general otherwise: this is what underlies the 'Heisenberg uncertainty principle' to the effect that non-commuting results cannot be observed simultaneously.) Instead of being specified as a mathematical measure, a 'complete theory' is now specified by some particular 'density' or 'Gibbs' operator, $\mathbf{P}$ say (with unit trace), on the Hilbert space. In the Heisenberg paradigm it is the contraction of this Gibbs operator with the projection operator, $\mathbf{X}$ $(= \mathbf{X}^2)$ say, of an observable result that specifies its probability $P = P(X)$, i.e. one uses the formula $P = \mathrm{tr}\,(\mathbf{PX})$.

Although the intrinsically 'probabilistic' approach ('indefensible as it is'[7]) is usually attributed to the mainly unpublished communications of the eighteenth-century thinker (one can hardly say writer) Thomas Bayes, its roots are older and its fully developed mathematical formulation and general (implicit or explicit) adoption more recent.[21] Even though most early scientists believed that empirical knowledge should 'in principle' be based on a deterministic treatment, the Bayesian approach to scientific analysis came to be widely adopted 'in practice', particularly when the conclusions were to be used as the basis for choosing a course of action (such as might follow, for example, from a medical diagnosis). The notion that the ideal embodied in the 'Determinist paradigm' might be dispensed with even 'in principle' began to emerge in the early days of statistical mechanics as pioneered by Boltzmann and Gibbs, but it was not until the rise of quantum theory, within living memory, that 'Determinism' was definitively displaced from its status as the generally recognised 'reigning paradigm' at the foundation of scientific thought. The irony is that (as is typically the case in a Kuhnian revolution) the paramount status of the 'Bayesian paradigm' was not clearly recognised until its own successor, the 'Heisenberg paradigm', had emerged on the scene.

In any of these (Deterministic, Bayesian or Heisenberg) paradigms, one is always free at any time, subject to the consistency requirements, to change the (binary, continuum measure, or Gibbs operator) weightings characterising one's overall 'theory' for any reason, whether desirable (e.g. because one has just discovered a new theory that is more economical in the sense to be discussed in the next section) or otherwise (e.g. because of inadvertent memory loss, due to, let us say, a computer breakdown). There is however a big qualitative difference in what happens when one acquires a new observational 'result'. In the Deterministic paradigm the new result is either consistent, in which case nothing happens at all, or not, in which case one's entire self-consistent 'theory' collapses completely, obliging one to think again from first principles. In the Bayesian paradigm one is required to have done more thinking about alternative hypotheses in advance, but as a result of this investment one is insured against being taken completely at a loss by a new observation, since its result can automatically be incorporated into the theory using the 'Bayes rule'. According to this obviously natural rule, the modified *a posteriori* theory is obtained from the original *a priori* version just by excluding the logical complement of the 'result' (where exclusion means deprivation of its status as a hypothesis by reduction of its measure to zero) and by adjusting the measure on the remaining part of the probability space (i.e. the part conditional on the 'result' statement) by a simple proportional rescaling so as to renormalise the total probability measure back to unity. In the Heisenberg paradigm, the analogous procedure is carried out by replacing the *a priori* Gibbs operator, $\mathbf{P} = (\mathbf{P}_0|\mathbf{X})$ obtained by analogously renormalising the corresponding projection of the *a priori* Gibbs operator, i.e. $\mathbf{P} = \mathbf{X}\mathbf{P}_0\mathbf{X}/\mathrm{tr}(\mathbf{X}\mathbf{P}_0)$. (This latter procedure, formalising what is commonly referred to as 'collapsing the wave packet', lies at the heart of an essentially metaphysical debate about the 'interpretation' of quantum theory, the dualistic 'Copenhagen interpretation' being, in so far as it makes any sense at all, less satisfactory from the Ockham razor point of view than the 'De Witt–Everett–Wheeler interpretation', or even than the 'solipsist interpretation' that might be preferred by 'one' who would rather have economy in the ontological sense referred to by Leslie[16] than in the Ockham sense to be discussed in the next section. There has also been much confused and embarrassed debate among classical probability theorists about the commuting Bayesian analogue of this question, but most classical theorists do not yet seem to have got round to realising the full depth of the problem nor the relevance of the commuting limit of the De Witt–Everett–Wheeler branching concept to their own epistemological discussions. In view of the 'Ossiander rule' to be discussed in more detail in sections 4 and 5, I shall not go any further into such perilous metaphysical arguments here.)

Before proceeding it is to be noted that anything describable within the Deterministic paradigm can also be expressed in Bayesian terminology, but not conversely, and similarly anything describable within the Bayesian paradigm can also be expressed in Heisenberg terminology, though again not conversely. For any particular application the requirement of consistency with a later paradigm in this series still leaves one free in practice to use the earliest one that will do the job. Despite the superior status of the rather sophisticated Heisenberg paradigm that is needed in microphysical applications, and the fact that the naive Determinist paradigm is still adequate for many practical purposes, it is the intermediate Bayesian paradigm that is most appropriate as a reasonable working compromise for use in a wide range of general scientific applications, including those with which we are concerned here. The traditional Bayes formula tells one explicitly how to apply the Bayes rule in any particular case so as to obtain the relevant *a posteriori* estimate of the appropriate 'probability' P for a particular 'hypothesis' of interest (e.g. the medical diagnosis of a particular disease) after taking account of the 'result' of an observation or test, starting from the corresponding (more primitive) *a priori* estimate, P_0 say. In practice this *a priori* quantity will often be rather precisely and objectively determinable on the basis of previous knowledge (in the medical example it might be derived from known comparative frequencies of the alternative diseases that might be suspect) but in general it must be considered as a subjective credibility weighting that each person has to decide for himself on the basis of his personal experience (or vestigial animal instincts) as best he can, the recommended 'scientific' method discussed in the next section providing guidelines that in many cases are frustratingly vague.

I find it convenient to convert the standard Bayes formula for doing such an adjustment to a mathematically equivalent version suggested on the one hand by traditional commercial usage and on the other hand by modern information theory. In this version one replaces the 'probability' measure associated with each hypothesis by a logarithmically related quantity that I call its (binary, decimal or Naperian) 'credibility rating' or simply 'credit', $C(P)$ say, which is defined (in terms of binary, decimal, or Naperian logarithms) by the expression

$$C = \log\left(\frac{P}{1-P}\right) \tag{2.1}$$

Whereas the probability measure P is limited to the range from 0 to 1, the corresponding credit C ranges from $-\infty$ to $+\infty$ (passing through the neutral credit zero when the probability is fifty per cent). The advantage of working with the 'credit' weighting (which is not a measure in the technical mathematical sense) rather than the 'probability' is that it allows the original Bayes formula to be conveniently recast in the very simple additive form

$$C = C_0 - D \tag{2.2}$$

where what I call the 'Debit', D, is a correction term which tells us how much the hypothesis is 'discredited' relative to its *a priori* credit value C_0 by the observational evidence. (The expression 'debit' may also be understood here as an abbreviation for 'deducted bits' if we use the standard (Shannon) information theory normalisation in which the logarithms are specified with base two.)

In order for one to be able to evaluate the debit D, the alternative hypotheses under consideration (whose *a priori*, and hence also *a posteriori*, probability weights are postulated to have been normalised so as to add up to unity) must each have a theoretical development that may be rudimentary but should at least be sufficiently developed to provide a corresponding provisional probability estimate for the observational results that are obtained. When transcribed into the present framework, the Bayesian prescription for the debit D associated with a particular hypothesis, Y say, is simply

$$D = -\log B \tag{2.3}$$

where the Bayes factor B is the ratio of the conditional probability for Y of the observed result, X say, to the complementary conditional probability of the *same* result (i.e. to the conditional probability obtained for X by averaging over the alternative hypotheses excluding Y). If there is only one alternative hypothesis, it can be seen that D will be completely independent of any assumptions about the *a priori* probability distribution, and that even if there are several the corresponding conditional probability will depend only on their relative but not on their absolute *a priori* probability weights, so that D will in this sense always be independent of C_0. It is to be remarked that successive measurements giving debits $D_1, D_2, \ldots$ will contribute additively to D in the form

$$D = D_1 + D_2 + \ldots \tag{2.4}$$

(i.e. the evidence will 'add up') provided the quantities measured are statistically independent, though not of course if they contain overlapping information. (It may also be remarked that the formulae (2.2) and (2.3) can of course be taken straight over into the Heisenberg paradigm of quantum theory provided the result X and the hypothesis Y correspond to projection operators that commute. In the non-commuting case the required factor B will still be definable, but its form will be more complicated than a direct ratio of conditional probabilities.)

Beyond the minimal requirement of providing an alternative conditional probability estimate, the alternative hypotheses need not be developed with the same degree of theoretical detail as the hypothesis Y in which one may be primarily interested, but there must be at least one vague alternative in order for a scientific discrimination procedure to have any meaning: the Bayesian probability weight P or the associated credit C of a scientific hypothesis has no absolute significance but takes its meaning only relative to a prescribed framework of comparison. (The apparently more objective, purportedly non-Bayesian procedures of statistical analysis developed by Fisher and others are merely ways of disguising the ultimately subjective *a priori* input that must always be present, using standardised prescriptions that are convenient when appropriate but which may be dangerous in the wrong context: like the anthropic principle to be described below, such prescriptions should be used with discretion and not blindly. It is also to be remarked that contrary to a common misbelief, one cannot obtain objectivity just by returning to determinism: ruling out fractional values and restricting P to be zero or unity is not a way of making it less subjective.)

The unattainability of absolute objectivity may be felt to be unsatisfactory, but in science (in the strict empirical sense in which the term is used here) some degree of subjectivity is inevitable because one is dealing, by definition (as has been reiterated by writers ranging from Francis Bacon to Pierre Duhem[20]), only in appearances – the study of absolute reality, whatever that may mean, being left to other branches of philosophy. (The acceptance of this defining limitation of empirical science as a subject does not of course prevent, though it may discourage, a scientist from personally indulging in metaphysical speculations when 'off duty'.) What has enabled us to establish the enormous body of (not absolutely but nevertheless relatively) 'objective' empirical knowledge possessed by our scientific civilisation today on the ultimately subjective Bayesian framework is the fact that when the theoretical models and observational results are sufficiently detailed and precise, the relevant 'debit' terms tend to be destabilised away from zero to very large positive or negative values, leading to conclusions, in the form of 'confirmation' or 'refutation', that are virtually independent of *a priori* input. The term 'confirmation' (with respect to the relevant Bayesian framework) is evidently to be understood here as meaning that the outcome is a high positive *a posteriori* credit value C (i.e. P very close to 1), while 'refutation' describes the opposite extreme for which one ends up with a large negative *a posteriori* credit value C (i.e. P very close to 0). The possibility of obtaining 'conclusive' evidence, as characterised by the condition that the 'discriminant' defined as the modulus of the debit should be very large, i.e. $|D| \gg 1$, is what makes the goal of scientific objectivity obtainable to a considerable degree in favourable circumstances. Such a favourable situation is to be contrasted with what may occur in a situation of strong 'prejudice' as characterised by $|C| \gg 1$ in which evidence that is short of overwhelming (i.e. for which the discriminant $|D|$ has a relatively moderate value) cannot affect the foregone conclusion, so that if different observers have incompatible prejudices (large C values with opposite signs) the controversy will remain scientifically unresolved. It is thus not irrational when someone (e.g. Dirac in the example discussed in section 5) who is strongly prejudiced takes little notice of new evidence: what is unwise is to allow oneself to be strongly prejudiced without valid and sufficient reason in the first place.

3. The Ockham Economy Principle

The considerations of the preceding paragraph are relevant to what is one of the most fundamental methodological problems in general scientific work, namely how to choose hypotheses deserving high credit ratings, and in particular how to decide what actually are appropriate *a priori* values for the credit ratings or the equivalent probability measures themselves.

Although the situation is far shakier in the other branches of philosophy and theology (where the ground is so unfirm that millions of man-hours of work have produced disproportionately little in the way of meaningful progress), the basis of 'natural' philosophy, meaning what is referred to here by the more popular name of 'science', is not as solid as its relatively spectacular progress (demonstrated by the recent rise of our high-technology civilisation) might lead one to suppose. Being concerned only with the superficial level of appearances (not with the deeper realities that other branches of philosophy and theology try so ineffectually to deal with), 'science' has the apparent advantage of disposing of large quantities of 'factual' data to work on. The trouble however is that any given finite amount of data will in principle be consistent with an infinite variety of theories, while even in practice (allowing for the finite size of our brains and computers) one can always extend the theoretical options far beyond what the data can determine.

Just as the standard procedure for using evidence for discriminating between hypotheses is associated in scientific folklore with the name of a rather obscure eighteenth-century clergyman, so, rather analogously, the standard guidelines for choosing hypotheses worthy of being taken seriously in the first place are traditionally associated with an even more remote (though in his own time more prominent) cleric whose 'razor' principle was propounded in the fourteenth century when science was still in an embryonic stage. I am sure that Ockham would be even more amazed than our own contemporary, Wigner,[22] if he could be informed of the truly miraculous success of the method associated with his name. (It is this very success that encourages the modern 'rationalism'[14] illustrated by the confidence of contemporary superstring enthusiasts in the motto that 'it is more important to have beauty in one's equations than for them to fit experiment'.[23]) In its primitive form, what in less colourfully metaphorical but more informative terminology may appropriately be described as an 'economy principle' is essentially just the astute precept to the effect that one should try to combine comprehensiveness with simplicity by giving *preference to hypotheses providing maximum relevant deductive output information from minimum independent input information*.

This Ockham principle constitutes the flimsy foundation on which our entire scientific edifice reposes. The 'miracle of science' is just how comprehensively the world we know by observation actually can be dealt with in terms of an input that is sufficiently limited to be easily processed by our limited human mental faculties. No contemporary of Ockham's could reasonably have guessed how much would be understood so soon using so little. Not only does the use of the 'economy principle' pay: it pays exhorbitantly and fast! Not only do we find ourselves mentally equipped to understand far more than might be considered as minimally necessary for the survival of our species: we have acquired a knowledge and mastery of nature that threatens to be the source of our collective destruction. The metaphor that 'nature is not perverse' understates what a theologian might express as the 'revelation' that 'God ... or the Devil? ... is on the side of the scientists'! (The ultimate reason for its own success is of course beyond the scope of science itself: such a question can only be addressed in the less modest context of theology.)

The Ockham principle acquires its greatest efficacy when the vagueness (or, to use the 'razor' metaphor, 'bluntness') of the formulation given above is reduced by the use of appropriate accessory principles as guides to its interpretation and application. As far as the ordinary scientific (as opposed, for example, to theological) applications with which we are concerned here are concerned, the main vehicle for the application of Ockham's economy principle is the use of the Bayesian paradigm (or, for microphysical applications, the Heisenberg paradigm) as described in the previous section. In the framework of the Bayesian paradigm the above formulation of the economy principle acquires a rather more specific interpretation in which the term *preference* is to be understood quantitatively (but in general by no means precisely) as meaning *assignation of a comparatively high* a priori *credit rating*. By enjoining us to give such preference, among theories with comparable predictive power, to those involving the smallest number of input parameters, this interpretation of the Ockham principle enables us to avoid or at least postpone the problem of attributing probability measure over the space of values of extra (one hopes superfluous) parameters in less simple theories by attributing comparatively negligible *a priori* credit to them at the outset. In the technical language of the preceding section, the Ockham principle conveniently 'prejudices' use against the less simple theories that we would in any case have more difficulty in setting up. (The scientific miracle referred to above is that we can so often get away so well with such wishful thinking.)

It would be tempting to try to pursue this line further by seeking a way to use the economy principle to specify a quantitative scientific value from which an appropriate *a priori* weighting might be calculated by some suitable formula, such an 'Ockham value' or 'productivity rating' being some sort of comparison of appropriate measures of 'production' (evaluating the deductive output) and 'investment' (evaluating the input information), the biggest difficulty being to decide what to consider as relevant. In order for a theory to be sufficiently complete to be testable, it is generally necessary to specify much more than just the simple or 'beautiful' mathematical model that may lie at its heart. (For example, the Schrödinger equation model at the heart of the both highly credible and highly productive non-relativistic quantum theory of the mechanics of light chemical molecules can be adequately described by one theoretical physicist to another on a few lines of paper (perhaps 10^4 bits of information); however, to explain the theory in a sufficiently developed form for a previously uninitiated layman to know how to set about testing it experimentally would require that one add on at least a fair-sized book as a technical appendix.) The way to quantify output information in a non-divergent manner is even less obvious, and there is also the question of the relative weighting of deductions of what is already known (which I think is mainly what people had in mind in the discussions of Ockham's time) and deductions of what can be but has not yet been tested, which (since the rise of experimental science) would be generally recognised as being of at least comparable value (a popular modern heresy, discussed in section 5, going to the extreme of suggesting that what is not yet tested is all that counts). Although it is thus very hard to see how to define 'absolute Ockham values', it should be rather easier to give meaning to some concept of 'relative Ockham values' of comparable competing theories, since one can hope for cancellation of the most ambiguous quantities involved: for the purpose of Bayesian discrimination this would be sufficient, since it is only relative weightings that matter (the absolute measures being normalised by the convention that the total be unity).

4. The Anthropic Selection Principle

The use of the anthropic selection principle can be construed as a way of interpreting the application of Ockham's 'razor' in a Bayesian framework when dealing with the interpretation of observational results that are correlated with the existence of the observer. This principle, for which a more exactly informative, albeit less poetically appealing, title would be the *observer self-selection principle* (since in principle it applies not just to ourselves but also to non-anthropic observers wherever they may exist in the universe) is based on the tautology that *an observational result that is inconsistent with the means by which it is observed has no chance of actually being observed*, whatever its Bayesian *a priori* 'probability' may be. It follows that in so far as the application of the Bayesian discrimination procedure is concerned we have no need to take any account of whatever probability measure may have been attributed to parameter values or hypotheses extending beyond what is 'cognisable', i.e. compatible with observability: by what one might call the 'anthropic theorem', *the* a posteriori *probabilities or credibility ratings will be independent of whatever* a priori *probability measure may have been assigned outside the range of compatibility with observation*. The crucial point on which this theorem depends is simply that the Bayesian debit D is just the difference of loga-

rithmic probabilities of the *same* (observed) result for alternative hypotheses:
the probabilities of other different results (observable or otherwise) are not
directly involved in the calculation.

This *anthropic theorem* means that instead of using more extended probabilities (if indeed they have been specified at all) we can just as well carry out
the entire Bayesian procedure from the outset in terms of conditional probabilities as restricted to parameter ranges and hypotheses compatible with observability: the corresponding conditional *a posteriori* probabilities or credit ratings will be the same as those deduced in a more extended framework. If one
needed to go over to the 'Heisenberg paradigm', the condition of 'cognisabil-
ity' would be represented by a certain projection operator, $\mathbf{C}$ say, that must not
only commute with but actually leave invariant the projection operator of any
observable result, X say, i.e. $\mathbf{CX} = \mathbf{XC} = \mathbf{X}$. The quantum version of the
'anthropic theorem' is the proposition that the *a posteriori* Gibbs density operator $\mathbf{P}$ will be the same whether calculated directly from an unconditional *a
priori* Gibbs operator $\mathbf{G}_0$ or from the corresponding conditional, observably
projected, Gibbs operator $\mathbf{P}_C = (\mathbf{P}_0|C) = \mathbf{CP}_0\mathbf{C}/\mathrm{tr}(\mathbf{CP}_0)$, i.e. we shall have $(\mathbf{P}_0|X)$
$= (\mathbf{P}_C|X)$. This means that for all empirical purposes it suffices to know the
'cognisable' Gibbs density operator $\mathbf{P}_C$, knowledge of a 'fundamental' unpro-
jected *a priori* Gibbs operator $\mathbf{P}_0$ being superfluous.

As is clear from the widespread unwillingness to accept it 'openly',[10] the
Bayesian paradigm described in section 2 involves more than the (rationally
indisputable) Bayes theorem of mathematical measure theory that lies at its
heart: the non-trivial point is the principle of associating quantitative 'probability' weightings to competing hypotheses in the first place. In an analogous
manner, the 'anthropic principle', even in its weak form, involves rather more
than just the 'anthropic theorem' enunciated in the previous paragraph. This is
why the attitude of someone (I have in mind the example of Dirac[13]) who
refuses to accept the weak anthropic principle need not be considered as irrational but merely as unempirical. (The prestige of great original geniuses such
as Newton, Einstein, or Dirac, is such that their devoted followers are liable to
be lamentably misled by their occasional lapses into the unscientific arrogance
of statements such as 'I don't make hypotheses', 'God does not play dice' or,
most pertinently to the present case, 'I prefer the theory that allows the possibility of endless life.'[13] It is partly to counter the dangerous examples set by
top-notch professional scientists of such heroic calibre that it is so important to
insist on the safer guidelines provided by the common-sense folklore doctrines
associated with more modest and less universally admired – legendary rather
than heroic – amateurs such as Ockham, Bayes or Malthus, whose lapses have
long been forgotten by all but professional historians, while the allegorical
version of their teaching stands, not on their reputation, but on its own merit.)

What makes the anthropic principle more than just a corollary of Bayes' theorem is the indication it provides as to how Ockham's principle should be interpreted in assigning the *a priori* probabilities that one requires. It is more than just the convenient elimination of any requirements of *a priori* probabilities for hypotheses and parameter ranges incompatible with observability. As was clearly implicit in the first applications suggested at the time of its formulation,[1,3,4,5] the essence of the (ordinary, i.c. weak) anthropic principle is contained in the following precept whose explicit formulation depends on the use of the technical concepts assembled in the two preceding sections. This rather more formal version of the anthropic selection principle enjoins that whenever a more adequately precise prescription based on previous knowledge is unavailable (i.e. 'we don't know something'[7]) as a guide, we should not start by trying to interpret and apply it at the level of an extended theoretical framework that may perhaps be definable beyond what is 'observable' or 'cognisable' instead the *anthropic principle* prescribes that *it is to the conditional probabilities as restricted by observability that the* a priori *preference for the most economic hypothesis should be attributed.*

Before continuing I would remark that, although it effectively reduces to a tautology in contexts for which the 'Determinist paradigm' is adequate, the 'anthropic principle' has a straightforward but non-trivial translation into the 'Heisenberg paradigm' as the principle that one should prefer the most economical theoretical prescription for the direct specification of the 'cognisable' Gibbs density operator P_C, rather than seeking simplicity at the supposedly deeper level of an unnecessarily extended unconditional *a priori* Gibbs operator P_0. This 'Heisenberg paradigm' formulation of the 'anthropic principle' is likely to be relevant to discussions of quantum cosmology as recently developed by Hartle and Hawking,[24] but for the applications with which I have been concerned so far, the Bayesian formulation in the previous paragraph is sufficient.

Having presented this explicitly Bayesian 'razor sharpener' formulation of what was always meant[1] to be understood by the (ordinary) anthropic principle, I would like to justify it by showing the unreasonability (if not actual irrationality) of refusing it due respect. If one has insufficient prior knowledge to fix the relevant a priori probabilities on an objectively firm basis, and if one does not accept the use of the anthropic principle as a means of limiting (if not completely eliminating) the ensuing subjective arbitrariness, then one condemns oneself implicitly to finding some other appropriate way of assigning an a priori probability measure over a theoretical framework extending beyond what is 'observable' or 'cognisable', the observationally relevant conditional probabilities being subsequently obtained (by projection) with values that need not automatically be compatible with what would have been given by the anthropic principle. What is hard to see in general is how the use of any such alternative procedure can be reconciled with the spirit of Ockham's economy principle. In their attempts to grapple with the enhanced degree of arbitrariness inherent in the use of a probability space extended beyond the minimal 'observable range', authors of the miscellaneous non-anthropic procedures that have been proposed on various occasions have had recourse to (e.g. aesthetic or even theological) considerations whose nature is, to say the least, hardly conducive to convergence of the Bayesian discrimination process towards an agreed conclusion.

As an illustration of an attempt to justify an 'extra-cognisable' *a priori* probability measure incompatible with the one got using the anthropic principle, I would refer to the classic example of the Dirac/Dicke controversy[12,13] that I have discussed in detail elsewhere.[25] In this example Dirac implicitly invoked a cosmological time weighted probability measure extending far beyond the part of the universe in which life such as ours can be imagined. Even if one takes account of the conceivable existence of radically different kinds of observers (such as Fred Hoyle's fictional black clouds) the appropriately modified anthropic probability measure would hardly be expected to agree with a weighting simply derived from cosmological time. Dirac justified his choice indirectly on the basis of a preference for a universe model in which life is eternally possible, a 'finality'[6] (as opposed to 'selection') criterion belonging to a philosophical or theological category that is rather beyond the scope of ordinary scientific methodology.

The 'Ossiander rule'[14] to the effect that one should draw no metaphysical implications from science ('Beware lest you expect truth!') was originally intended for protecting scientists from external interference. (It was Galileo's own rash violation of this rule that got him into trouble.) However, for a modern scientist an equally if not more important reason for steering clear of metaphysics is to protect his own scientific objectivity. In Einstein's case it was literally the ideal of 'truth' in the deterministic sense that lured him astray, but other ideals such as 'justice' can, for example in social science, be even more dangerously misleading as indicators of *a priori* credibility, however relevant they may be to a choice of action after the scientific analysis is complete. Moralistic eyebrows might be raised at the precept that, for a scientist, the most innocuous ideal is 'beauty', but a puritan may be reassured that this is only on condition that the aesthetic criteria are sufficiently austere to escape the economy 'razor'. Dirac followed 'beauty' (in the equations) all his life and got away with it: it would appear to have been the yen for 'immortality' that got him into trouble. Motives such as the currently widespread desire to establish that 'we are not alone' or that our universe can support 'endless life' are not a scientifically sound basis for attribution of *a priori* probability. Aesthetic preference is a much less misleading guide if one's ideal of beauty is sufficiently classical, though it would be treacherous if one's tastes were more baroque. So long as Dirac was guided by the 'aesthetic principle', which for someone of his taste had virtually the same implications as the economy principle, his predictions, such as the existence of the positron, were brilliantly confirmed. However his luck (in the Bayesian sense) deserted him when he let himself be guided by a metaphysical 'finality' principle antithetical to the razor criterion on which the anthropic selection principle is based: the observational evidence against Dirac's hypothesis of varying gravitational coupling has by now become so conclusive (in the sense of an enormous Bayesian debit, $D \gg 1$) as to overwhelm even the most strongly favourable *a priori* prejudice.

5. Refutability and the Strong Anthropic Postulate

In view of the frequency with which it is invoked when the foundations of the scientific method are in question, particularly from a philosophical point of view, I feel obliged to digress at this stage to comment on a popular but in my view deviant and misleading variant of the economy principle that is commonly associated with the name of Popper. (In so far as I refer with approbation or reservation to the methodology of Ockham and Bayes on the one hand or Popper on the other, I am not concerned here with the historical question of what any of these individuals actually thought or taught, but merely with the folklore doctrines named after them allegorically.) According to the currently popular version of 'Popper's razor', the scientific respectability of a theory (subject of course to logical self-consistency) is essentially dependent on the paradoxical requirement that it provide *refutable predictions*, whereas according to a common-sense idea of empirical scientific orthodoxy on the basis of the economy principle, the minimal respectability requirement would be provision of *confirmed deductions* (without which the theory would be taken seriously only by an extreme 'rationalist'[14]). As for further *testable predictions*, although not obligatory for respectability, their provision would of course in general be considered a bonus. It would not even be a bonus however if (as can occur when none of the likely 'debits' is strongly negative) there are no confirmable but only refutable predictions. As an attempt to indoctrinate innocent scientists with the notion of the indispensability of something that need not even be intrinsically advantageous, 'Popper's razor' is patently malicious (even if nature is not): it may impress gullible theoreticians but it cuts no ice with hard-headed engineers.

On closer scrutiny, one of the points of view from which the Popper 'refutability principle' can be objected to is that it implicitly violates two different kinds of formal symmetry that are more or less inherent to the Bayes–Ockham methodology on which empirical science as I understand it is effectively based. The first of these is the formal symmetry between prediction and retrodiction that is violated by the suggestion that the former is to be preferred, whereas as far as the economy principle is concerned the latter is at least equally valuable. As far as the Bayesian discrimination procedure is concerned, an untested prediction is of course better than a deduction that has failed a test, but it is absurd to suggest it is better than a deduction that has already been confirmed. Finally, as far as practical utility is concerned, the prediction of something new is not the only thing for which a theory may be valuable: there are many situations (e.g. in advanced military software) where it may be of literally vital importance to be able to replace a cumbersome tabulation using a lot of memory by a simple, rapidly executable, algorithm.

A more seriously (since less obviously) misleading symmetry violation in the popular version of the 'Popper principle' is the implication that refutation is somehow on a different and sounder footing than confirmation. It is of course quite valid to say that theories can be refuted but never absolutely confirmed, but the key to a correct understanding of this statement is to notice the location of the qualification 'absolutely'. It is not only equally valid, but actually equivalent, to make the complementary statement (which must be comforting for our contemporary superstring enthusiasts) that theories can be confirmed but never absolutely refuted.[20] (The impossibility of absolute confirmation includes the implication that a refutation can never be absolutely confirmed.)

I suspect that the negativist insistence on refutation rather than confirmation may in part reflect a 'sour grapes' attitude on the part of philosophers of other branches who subconsciously envy the positive achievements of their scientific colleagues. The penetration of negativistic attitudes within the scientific community itself may also reflect ambivalence and frustration among its own more theoretically minded members who are evidently the most susceptible to the temptation of 'rationalism',[14] tending to yearn for the ideal of an ultimate 'complete theory of everything' having as its heart an elegant mathematical model (e.g. a Hartle–Hawking type cosmology[24] based on a superstring theory), but knowing that (even if the essential nature of such a model could be guessed) the theory would still be beyond hope of firm (never mind absolute) empirical verification in view of the many supplementary hypotheses that would be needed to tie it in (presumably via the 'cognisable' density operator $\mathbf{P}_C$ mentioned in the previous section) with what we actually observe.

An example of this academically fashionable negativistic point of view is the statement that Newton's theory of solar system dynamics has been 'refuted' by the discovery of deviations at the level of nearly one part in 10^5.

This is to be contrasted with the more positive attitude that prevails in engineering circles where one would describe the same situation by the statement that Newton's theory of solar system dynamics has been 'confirmed' to be reliable to within better than one part in 10^4. From a rigorous Bayesian point of view each of these (consistent but non-equivalent) statements is equally respectable: neither is absolutely certain (nor absolutely well defined) but both may be considered to have a high degree of scientific credibility. Note that both are still in principle (but not absolutely) refutable and that this is not a bonus but a weakness: it is always just conceivable that some Newtonian mechanism (e.g. involving the hypothetical planet 'Vulcan') may after all account for the one in 10^6 deviations that one usually attributes to relativity, while on the other hand (as is seriously hoped by the 'fifth force fanatics') some as yet hidden non-Newtonian effect may eventually show up at the one in 10^3 level.

By unfairly conveying the impression that confirmation is valueless unless absolute, without also subjecting refutation to any such unreasonably idealistic requirement, the folklore version of the Popper principle effectively reduces both confirmation and refutation to meaninglessness. In such a tendentious system, anything qualifying as 'scientific' is not only unconfirmable but also, sooner or later, refuted, so that as far as the scientific practitioner is concerned the Popper principle is not so much fallacious as, in the end, effectively empty. (Being in Venice, one can hardly fail to recall the Gilbert and Sullivan aphorism that when everybody is somebody then no-one is anybody.) By such biased means (and this I suspect is the subconscious motivation of the whole exercise) is natural philosophy unjustly made to seem as ineffectual as the other branches.

The essential logical symmetry between the useful notions of (relative, not absolute) confirmability and refutability in a Bayesian framework is apparent from the consideration that refutation of a hypothesis is equivalent to confirmation of the complementary hypothesis specifiable as the appropriately weighted combination of the various alternatives. This symmetry is just the reflection of the invariance of the formalism described in section 2 under the simultaneous sign changes

$$C \leftrightarrow -C, \; D \leftrightarrow -D \tag{5.1}$$

of credit and debit (corresponding in the more conventional language or ordinary probability to $P \leftrightarrow (1-P), B \leftrightarrow B^{-1}$) expressing interchange of the given hypothesis under consideration with its Bayesian complement.

Having pointed out the absurdity or vacuity of insisting on (absolute or relative) refutability as a (necessary or sufficient) criterion for scientific 'respectability' I would emphasize again that this is in no way to deny that observationally testable, and therefore not only (relatively) confirmable but in general also (relatively) refutable, predictions should nevertheless be considered as a positive contribution to a theory's scientific 'value' as loosely defined in section 3.

My attention was drawn to the misuse of 'refutability' as a criterion for respectability when the anthropic principle was subjected to criticism from this point of view. However, my insistence that 'refutability' is not intrinsically an advantage should not be read as suggesting a claim of 'irrefutability' even for the anthropic principle itself, and still less for its diverse and often dubious applications. Many of the predictions that have been derived using the anthropic principle are not only 'refutable' but in real danger of actual 'refutation' (i.e. being highly discredited in the Bayesian sense defined in section 2). In particular, such vulnerability applies even to what I originally called the 'strong anthropic principle' (though, as John Barrow has remarked, the term strong anthropic 'postulate' would be more appropriate), meaning the combination of the ordinary weak anthropic principle with a hypothesis of the existence of an ensemble of connected or disconnected branches of the universe over which 'fundamental constants' would have an extended range of values. In such an ensemble the familiar observed values of the parameters would be interpretable as deriving from anthropic selection. I had been inclined to think that this recently much discussed postulate[6, 26] was exposed to the objection that the invocation of all these extra branches of the universe was not very economical in the sense required by Ockham's principle, but Dennis Sciama[27] has presented a seductive argument to the contrary in one of the accompanying discussions. An objection of a superficially similar but fundamentally different kind that has been made against multiply branched cosmology (though more *apropos* of the De Witt–Everett–Wheeler interpretation of quantum theory than of the strong anthropic principle by itself) is that of 'bloated ontology'.[28] According to the rules of the scientific game however this is not quite fair since, as has been generally agreed from Francis Bacon's time down to our own (see for example the writing of Pierre Duhem[20]), questions of ontology lie outside the scope of empirical science as such. The trouble is that although it is easy to preach as Dirac did that (in accordance with the Ossiander rule referred to in the previous section) one should not worry about such 'type 1' questions,[23] in practice no-one can constrain himself to being a 'true scientist' for twenty-four hours a day, and many have followed Dirac's own unfortunate example (of forgetting the Ossiander rule) by succumbing to the temptation of letting their judgement be swayed by metaphysical extrapolations.

What I specially want to do here is not to defend the strong anthropic postulate but on the contrary to point out that the consideration that the other branches of the universe would (certainly if disconnected) not be directly observable does not mean that this strong anthropic postulate is irrefutable when taken as a whole. It might in fact be refuted by our finding out (e.g. by the discovery of something like Fred Hoyle's fictional black cloud) that systems of observers could after all exist under conditions quite different from those that seem to be necessary for our own kind of life. Moreover, like Dirac[23] himself, many theoretical physicists are optimistic that it may soon be possible to undermine the theoretical basis for the strong anthropic principle by discovery of underlying physical mechanisms fixing what, at our present level of understanding, appear to be independent fundamental constants, so that at the new deeper theoretical level other values would cease to be even conceivable, which would however still leave a mystery (that I would call the 'anthropic enigma') if it were possible (contrary to the eventuality suggested above) to confirm the impression that finely tuned values are needed for observers to exist. The very real vulnerability of the strong anthropic postulate to being non-viable in such a way may make it seem more scientifically respectable in the eyes of believers in the 'Popper principle', but it makes it less convincing for me.

Of course when a particular application of the anthropic principle is 'refuted', it is the other hypotheses involved in the application that one would be inclined to reject first before casting doubt on the general anthropic principle itself. (In the above example it is the 'strong' postulate of different fundamental parameter values in different branches of the universe that would be sacrificed before one abandoned the 'weak' anthropic principle itself.) Nevertheless it is always conceivable that if too many anomalies were to turn up systematically, not only the anthropic principle, but even (as Hoyle and Wheeler among others[7] have remarked) such a firmly founded general statistical principle as Boltzmann's law of entropy increase, could at some level of precision be cast into doubt, with implications (depending on the type of anomaly) that would be, to say the least, fascinating in either case.

6. Ultra-Darwinian Selection

The application[2] to be described in the remaining sections is not just an illustration of the (weak) anthropic selection principle but something of considerable interest in its own right. Its purpose (achieved using the simple 'lock sequence model' described below for the overall process) is to throw some new light on the way the ecologically stochastic Darwinian evolution mechanism based on the Wallace–Darwin principle of natural selection could have given rise to the existence of our own, now civilised, species on this planet. Despite the progress brought about by the genetic theory of Mendel and Morgan and the more recent discovery of the DNA coding mechanism, there remain vast gaps in our understanding of the way the Darwinian process works out in detail, even at the level of individual species, and still more at the level of the highly complicated ecological system constituted by the interaction of all the species with each other (and with the geophysical environment) on a planetary scale. One of the simplest questions one can ask is about the theoretically expected timescales governing the stochastic process of evolution of salient developments, such as the DNA coding mechanism itself, or the multicellular nervous system of the more advanced animals but not plants. All one can do for such a process at a theoretical level is to place a very conservative lower limit, which must at least correspond to a moderately large integer multiple of the relevant breeding generation timescale τ_g. However such conservative estimates[2] are extremely short, particularly for microorganisms, compared with what, in view of the extreme complexity of the genetic and ecological processes involved, one might expect, and also compared with what is actually observed to have occurred in the fossil records.

As a very crude mathematical model (in view of our ignorance nothing more elaborate would at present be justified) with which to investigate what we can find out observationally about the characteristic timescales, it seems reasonable[2] to envisage a stochastic process rather analogous to the process of getting through a sequence of doors with combination locks of assorted difficulty corresponding to the complexity not just of the genetic but also the ecological processes involved at each stage. In this process one is restricted to an unsystematic random method in which one does not even keep note of what has already been unsuccessfully tried. On timescales large compared with what is needed to try a single combination, which in the biological analogue corresponds to at least a moderately large multiple of the breeding timescale τ_g, such a process can be modelled by a continuous probability distribution dP/dt for the time, t say, of ultimate arrival, where $P(t)$ is the probability of arriving any time before the value t. If there were only a single lock, $P(t)$ would just be given by an expression of the simple form $1 - \exp(-t/\bar{t})$ where $\bar{t}$ is a characteristic mean timescale (corresponding to the product of τ_g with a numerical factor proportional to the degree of complexity of the process, which may be very large). For a large number of locks with very diverse individual characteristic timescales, the resulting formula for $P(t)$ will be more complicated but qualitatively it will still be a function that rises monotonically from zero to a saturation value that may be less than unity if we allow for the not unrealistic possibility that some or even most combinations may open alternative doors leading to other bifurcating (in our case macroecological) pathways incompatible with the outcome in which we are interested (and perhaps in some cases leading to autodestruction of the system itself). For a single lock, dP/dt decreases monotonically from the outset, but for several locks it will start very small and increase to a peak value at a time comparable with the overall characteristic time, $\bar{t}$ say, at which P first reaches a value comparable with the saturation value. For higher t values, as P approaches saturation, dP/dt must of course decrease again to zero as in the single lock case.

In an observational study of such a system the first obvious thing one might hope to estimate would be the overall characteristic timescale $\bar{t}$. With more detailed observational results one might hope to find out something about the separate timescales of the individual locks involved, but to do this one must evidently develop the mathematics a little further. Even if we had a large sample of observations allowing us to map out the distribution $P(t)$ completely, it would be hard to learn much from the flat part for $t \gg \bar{t}$. However, the lower part of the range can be more informative, since in the neighbourhood of a particular value of t in the lower range, $\tau_g \ll t \ll \bar{t}$, the distribution may be expected to have an approximate power law behaviour of the form

$$P(t) \simeq a\, t^n \qquad (6.1)$$

where the index n is an integer that can be interpreted as estimating the number of locks or evolution stages that are 'critical' with respect to the timescale t under consideration, i.e. the number whose separate individual timescales are long compared with that value of t.

The kind of 'lock sequence model' that has just been described[2] may be appropriate for representing the crude overall stochastic comportment of the long-term Darwinian evolution of macroecological systems of many kinds on various scales. It is to be noted that short sequences of locks or evolution stages may be strung together to make longer sequences of the same kind. Our purpose here is to apply this kind of simplified model to the longest such sequence that we know about, namely that of the various stages by which our global planetary ecosystem reached the stage at which our own civilisation appeared.

Before continuing I would like to emphasise the importance of an anthropic selection effect of an intrinsic kind that occurs when one is limited to observing one's own evolutionary past, and that is essentially distinct from the anthropic selection mechanism of extrinsic astronomical origin that will be discussed in the following section. The intrinsic selection effect, on which it seems harder to get an observational handle, is the one resulting from the bifurcating macroecological branch pathways, incompatible with a scientific civilisation as the outcome, whose likely existence was referred to above as a reason for the distribution $P(t)$ to level off at a maximum small compared with unity. The idea that such unfruitful pathways may be much more probable *a priori* than pathways that lead to the existence of scientific observers may be considered to be supported to some extent by the kind of considerations that have been used as an argument against purely (neo)Darwinian theories in favour of 'design' mechanisms of a teleological nature.

Supporters of the more scientifically acceptable class of purely causal theories have traditionally tended to argue that the appearance of teleological purpose (too systematically directed towards our own emergence to be explicable by the random action of blind Darwinian natural selection) is merely an illusion. Although many of the purported examples of manifest purpose almost certainly are quite illusory, it is not quite so easy to dismiss them all. What I want to point out[2,25] is that there is no need to dismiss them all: the existence of apparent purpose in the form of systematic direction towards our own emergence is not merely compatible with causality but actually to be expected if one takes account of the intrinsic anthropic selection effect.

Just as (almost) everyone agrees that the apparently purposeful growth of an acorn into an oak tree can in principle be accounted for without recourse to teleology in terms of ordinary Darwinian natural selection mechanisms, so, on a larger scale, the apparently purposeful evolution of the planetary macroecological life-system from one limited to microorganisms to one that includes the civilisation we have today is analogously explicable in principle, without recourse to design arguments, in terms of what might be called ultra-Darwinian selection, meaning the combination of ordinary Darwinian natural selection (preferring certain genetic strains to others) on a local scale with anthropic selection (preferring planets on which observers ultimately emerge to those on which evolution gets sidetracked or blocked) on a galactic or cosmological scale.

If this ultra-Darwinian (natural plus anthropic) selection picture is correct, it means that the typical behaviour of local Darwinian natural selection in other planetary life systems that we might hope one day to observe would be very different from what we trace in the fossil record on our own planet, whose status would be that of an anthropically biased sample in which there might indeed be many apparent 'design' effects that would be absent on the others.

The effect of the intrinsic anthropic selection (elimination of macroecological evolutionary sidetracks) that we have been considering here entails, if it is important, the implication that extraterrestrial civilisations will be much less common than if the postulated bifurcation possibilities did not exist. In the next section I shall discuss a quite distinct extrinsic anthropic selection effect that will be seen to imply that our terrestrial specimen is probably biased not only in the direction but also the rate of the Darwinian process. The extrinsic anthropic selection effect also leads independently of the internal one to a prediction of comparative rarity of extraterrestrial civilisations. Since as far as this aspect is concerned the intrinsic and extrinsic selection effects are cumulative, my own guess (though it is hard to be precisely quantitative) is that extraterrestrial civilisations are likely to be very rare indeed, to the extent that the nearest may be outside our galaxy.

7. The External Cutoff

In order to subject a theoretical model, such as the 'lock sequence model' described in the previous section, to observational testing, one must, as explained in section 2, have some *a priori* idea of the relative probabilities of the relevant parameters. What we have in mind here is the application to long-term macroecological evolution on a planetary scale in a steady external astronomical environment, for which the relevant parameters should presumably just be functions of the initial geophysical conditions on the planet.

Before proceeding, it is to be noted that even on the assumption that the external astronomical environment does not undergo drastic variation, the geophysical environment may undergo major variations caused by the ecosystem itself as it evolves through various stages. The corresponding geophysical response timescales, in addition to the breeding timescales already referred to, must be considered as determining a lower cutoff to the timescales on which the simple form of the 'lock sequence model' is applicable. In his accompanying discussion, Jean Heidmann[29] has raised the question of whether the geophysical response timescales might not be so long as to impose the necessity of modifying the model in order for it to be applicable to our own case, but since such a necessity is far from being clearly established I shall apply the Ockham principle by continuing here on the basis of the provisional assumption that the simplest model is adequate.

Although the parameters characterising the model should in principle be calculable as functions of the initial geophysical conditions, since we are in practice quite unable to give any theoretical estimates (beyond the condition that they greatly exceed breeding timescales) for the characteristic timescales associated with various stages, and in particular for the characteristic mean timescale $\bar{t}$ for the entire stochastic process to a stage of development comparable with our own, we can only proceed on the basis of observation. However, as is typically the case when one starts from scratch in a new experimental field in which stochastic models are appropriate, one immediately stumbles on the problem that, as emphasised in section 2, one needs *a priori* estimates before one can draw any conclusions at all from any observational results that may be available, and in the last resort one only has Ockham's not always very precisely interpretable principle to guide one.

For the case in hand, that of evolution to our own stage of development, we are handicapped by the fact that (since the search described in Heidmann's accompanying discussion[29] has not yet shown any signs of positive results) the only observational result available is the time t taken for evolution in our own case. Since this is a self-observation it is evidently subject to the anthropic self-selection principle, at least in its most rudimentary form as the motto 'observer beware'. It can in fact be seen that there is indeed an important selection effect (distinct from the intrinsic selection effect mentioned at the end of the previous section) that must be taken into account in the interpretation of the observed evolution time t. Treating this effect in the manner described in section 4, we can take advantage of the anthropic principle as a means of sharpening Ockham's principle in such a way as to obtain useful restrictions on the arbitrariness of the *a priori* probabilities which enable us to draw at least moderately convincing and certainly interesting conclusions.

The relevant selection effect results from the fact that the reasonably steady external astronomical environment needed for the continuation of the evolutionary process can never in practice last indefinitely. More particularly, in the case when the environment is maintained by a hydrogen-burning star like our sun, the process described by the simple lock sequence model (or any more sophisticated variant thereof) will inevitably be brutally interrupted when the star reaches the end of its main-sequence life after a time τ_0 say. This means that as far as observational testing is concerned (and this applies not only to self-observation but to any similar extraterrestrial systems we may one day be able to observe) the relevant probability distribution is not what is obtained directly, e.g. from the simple lock sequence model, on the assumption of a permanently steady external background, but a truncated modification retaining its original 'steady background' form for $t < \tau_0$ but with dP/dt dropping abruptly to zero so that P remains constant beyond.

If we now restrict our attention to the case of self-observation, it can be seen that the truncation effect will entail a selection effect of the anthropic type: only the part of the time range below the cut off point τ_0 will be observable. On a planet where the evolutionary process had not reached the stage of emergence of a scientific civilisation before the external cut off time, no self-observation could be made at all. Therefore according to the anthropically sharpened version of Ockham's principle (section 4) the attribution of comparable *a priori* probability measures to comparably simple hypotheses should be applied at the level of the *conditional probabilities* as restricted by the cut off condition $t \leq \tau_0$.

If one accepts the weak anthropic principle in this form, together with the hypothesis that the truncated version of the lock sequence model is appropriate, then the remainder of the argument that follows is essentially though not quite exclusively deductive. What we want to do is to learn as much as we can, starting with an estimate of the overall mean timescale $\bar{t}$ characterising the relevant 'steady background' (untruncated) distribution as modelled in the previous section using just the single observed value of the time t taken for our own evolution since the stage when steady thermally and chemically favourable conditions were established on earth.

In a typical precision time measurement in the laboratory one usually supposes that the probability distribution is normal with respect to a peak value $\bar{t}$ (that one wants to know) whose *a priori* probability distribution is assumed to be approximately flat over a range large compared with the standard deviation, the latter being considered as a known function of the timing apparatus. In this familiar kind of situation the Bayes procedure leads of course to an *a posteriori* probability distribution for $\bar{t}$ that is also approximately normal (with the instrumental standard deviation) with peak at the observed value t.

The case with which we are concerned here is technically more complicated and its conclusion more sensitive to what is assumed *a priori*, but the underlying principles are much the same. The probability distribution given by the truncated lock sequence model is very far from the normal type, but, in view of our lack of information suggesting anything else, it will still be reasonable to suppose that the characteristic time parameter $\bar{t}$ has a fairly flat *a priori* probability distribution, most plausibly with respect to logarithmic rather than linear scale, over the range of interest extending from the lower cut off value given by some multiple of an appropriate breeding timescale τ_g to values that, in view of the extreme complexity of the processes involved, can easily be imagined to be extremely large compared with the observed value of t. Unlike that which occurs for the normally distributed laboratory measurement, the form of the *a priori* distribution over the higher part of the range of $\bar{t}$ (i.e. whether the scale with respect to which it is 'fairly flat' is logarithmic or linear, and where it is ultimately cut off) is not without influence on the conclusions that one can draw, and as we know so little about it there is no point in attempting a high resolution analysis.

The most that is reasonable in these circumstances is a crude treatment in which the range is divided roughly into two bins, the first given by $\tau_g \ll \bar{t} \ll \tau_0$ and the second, to which it is reasonable to accord at least comparable *a priori* probability, by $\tau_0 \ll \bar{t}$ (It is evidently reasonable to accord comparatively negligible *a priori* probability to the borderline range $\bar{t} \approx \tau_0$.) The crucial point where the anthropic principle comes into the analysis is in specifying that these *a priori* assignations are to be understood as applying to the conditional probabilities subject to the restriction that t must lie in the observable range $t \leq \tau_0$. If they were applied to the unconditional probabilities, which would be unreasonable but not logically impossible, then the *a posteriori* conclusion would be radically modified.

8. The Implications of the Evidence: a Surprise and a Warning

We are now at last ready to take account of the actual observed result which is well-known from the geological record as being given with sufficient (better than fifty per cent) precision for our purpose by $t \approx 5 \times 10^9$ years. This leads to a very clear-cut conclusion[2] since it is remarkably close in magnitude to the almost equally well-known main-sequence lifetime of the sun, which is predicted on the basis of thermonuclear fusion theory to be given by $\tau_0 \oplus 10^{10}$ years, their relation being expressible by

$$\tau_0 \approx 2t \tag{8.1}$$

This coincidence of two such intrinsically different time estimates to within a factor of order two has attracted much less attention from theorists than it deserves, despite the fact that it is much more precise than the equally long known and much more crudely approximate (hardly within two powers of ten) order of magnitude coincidence that inspired the Dirac/Dicke controversy referred to above.

From the analysis of the previous section, it can be seen that (8.1) is in fact interpretable (in the terminology of sections 2 and 4) as signifying that there is a large anthropically conditional Bayesian debit, $D \gg 1$, for the first hypothesis, $\bar{t} \ll \tau_0$, for which the observational prediction is $t \ll \tau_0$ with high conditional (and also absolute) probability. The observed result (8.1) is evidently much more consistent with the alternative hypothesis $\tau_0 \ll \bar{t}$, predicting only the weaker inequality $t \lesssim \tau_0$ with high conditional (but not absolute) probability. This latter hypothesis therefore gets a high *a posteriori* credit rating unless its anthropically conditional *a priori* credit rating is very low compared with what seems reasonable. In short the evidence argues rather strongly in favour of the conclusion

$$\bar{t} \gg \tau_0 \tag{8.2}$$

In so far as this conclusion is merely a reinforcement of what was already my *a priori* opinion, making no contribution to solving the, for me far more interesting, question of by how great an order of magnitude the characteristic time $\bar{t}$ is greater than τ_0, I found this result somewhat disappointing. As discussed in the accompanying article by Jean Heidmann,[29] a searcher for extraterrestrial civilisations might be inclined to find it disappointing for the opposite reason, namely its evident implication that they are much more rare than one might have hoped (just how much so remaining unclear).

However, a more complete analysis leads to a further conclusion that is much more exciting, undermining what had been my *a priori* prejudice, and providing perhaps a little gleam of encouragement to the extraterrestrial search programme. Although our single time observation provides less information about $\bar{t}$ than if we had been dealing with a normally distributed variable in a conventional precision time measurement, we can instead obtain information of perhaps even greater interest about the number n say of distinct stages (as represented in the analogy of section 6 by distinct combination locks) that were 'critical' for our existence in the sense of having separate characteristic times large compared with the observed time value as given by (8.1). It can be seen from (8.1) that when (8.2) is satisfied the truncated probability distribution given by the lock sequence model will have a peak, not in the neighbourhood of t as in the familiar normally distributed case of conventional laboratory experiments, but at a value given instead by the approximate formula

$$t \simeq \left(\frac{n}{n+1} \right) \tau_0 \qquad (8.3)$$

Combining this with the approximate formula (8.1) for the observed result, one can immediately read out the 'measured value' of n as being of order unity (one or perhaps two but not much higher). (In terms of the lock sequence terminology this means that not much more than one or two of the combination locks are really difficult. This might be thought to favour the chances of more frequent occurrence of extraterrestrial civilisations, but it should be borne in mind that a small number of extremely difficult locks could be harder to get through than a larger number of easier ones.)

Of course, any naive 'deduction' from observation, such as that of the previous paragraph, is in fact a disguised Bayesian induction. In a more rigorous analysis, what one should say is that relative to the hypothesis that n is of order unity (one, perhaps two but not much higher), there is a very large Bayesian debit, $D \gg 1$, for the alternative hypothesis that n has some other value (equal to zero or exceeding three, since the number of critical stages is intrinsically non-negative). It is thus not objectively obligatory to assign a high *a posteriori* credit rating C to the conclusion that n is of order unity. Nevertheless to avoid doing so one would need to have a very high *a priori* prejudice in favour of higher values of n. I emphasise this element of doubt because, for reasons mentioned in my original discussion[2] of this question, I did in fact have an *a priori* prejudice in favour of considerably higher values for the number n of critical stages (while, for reasons that I find less convincing,[2,25] Barrow and Tipler[6] have argued in favour of even more extreme values). My own prejudice (I have no authority to speak for them) was, however, by no means strong enough to outweigh the evidence described above, which I believe should be taken seriously.

It is nevertheless to be pointed out that even if one's prejudice is insufficient to outweigh the above evidence within the framework of the assumed model, there is still the possibility of calling into question the assumptions on which the model itself is based. In the foregoing discussion one assumption that is perhaps more vulnerable than the others is the supposition that the external astronomical cut off time τ_0 that is relevant in our own case should be the entire main-sequence lifetime of the sun. Our understanding of the global working of the coupled system of the earth's crust, oceans, and atmosphere is insufficient to rule out the possibility of envisaging that the very slow heating up of the sun, as it evolves within the main sequence, might not have been destined to trigger an instability that would have rendered our environment uninhabitable after a time τ_0 significantly shorter than needed to reach the stage of much more rapid and drastic heating at the end of the main sequence era.

If this not very far fetched alternative supposition is correct the observations would need to be reinterpreted as implying a larger value of n, while conversely the arguments that originally inclined me to believe in a higher value of the number n of critical steps could be interpreted as suggesting that some kind of atmospheric destabilisation was imminent when (by what would have been a stroke of luck within the restricted context of this particular planet, but an anthropic selection effect from a more cosmological point of view) our evolution reached the stage of scientific civilisation while it was still just, but only just, possible.

This latter alternative interpretation of the evidence implies, as a corollary, a rather serious warning. If our atmosphere was in danger of external destabilisation within an astronomically short time, this would seem to imply a much more immediately imminent danger of destabilisation by the action of our own civilisation: collectively we have already shown ourselves to be capable of causing larger atmospheric modifications within the timescale of a human lifetime than the slow main-sequence evolution of the sun could have brought about in millions of years.

Acknowledgements

The author wishes to thank the organisers and participants of the 2nd Venice Conference on Cosmology and Philosophy, and most particularly John Barrow, George Ellis, Jean Heidmann, Hubert Reeves, and Dennis Sciama, for critical discussions and constructive suggestions.

References

1.   B. Carter, Large Number Coincidences and the Anthropic Principle in Cosmology, in *Proc. 1973 I.A.U. Symposium on Confrontation of Cosmological Theories with Observational Data*, ed. M.S. Longair (Riedel, Dordrecht, 1974).

2. B. Carter, The Anthropic Principle and its Implications for Biological Evolution, in *Proc. 1983 R. Soc. Discussion Meeting on the Constants of Physics,* ed. W.H. McRea and M.J. Rees (R. Soc., London, 1983) and in *Phil .Trans. R. Soc. Lond.,* **A310**, 347 (1983).

3. C.B. Collins, S.W. Hawking, *Astroph. J.* **180**, 317 (1973).

4. G.F.R. Ellis, *J. Gen. Rel. and Grav.,* **9**, 87 (1978).

5. B.J. Carr, M.J. Rees, The Anthropic Principle and the Structure of the Physical World, *Nature*, **278**, 605 (1979).

6. J.D. Barrow, F.J. Tipler, *The Anthropic Cosmological Principle* (Oxford University Press, 1982).

7. Mr X, in *The Nature of Time*, ed. T. Gold, D.L. Schumacher, p. 110 (Cornell University Press, 1967).

8. J.M. Keynes, *A Treatise on Probability* (London 1921).

9. P.S. Laplace, *A Philosophical Essay on Probabilities*, trans. F.W. Truscott, F.L. Emory (Dover, New York, 1851).

10. J. Madox, New Twist for Anthropic Principle, *Nature*, **307**, 409 (1984).

11. A.D. Linde, *Phys. Lett.*, **B129**, 127 (1983).

12. R.H. Dicke, *Nature*, **192**, 440 (1961).

13. P.A.M. Dirac, *Nature*, **192**, 441 (1969).

14. G.Gale, *Theory of Science: an Introduction to the History, Logic and Philosophy of Science* (McGraw Hill, New York, 1979).

15. H. Pagels, A Cozy Cosmology, *The Sciences* (March/April, 1985).

16. J. Leslie, The Scientific Weight of Anthropic and Teleological Principles, *Proc. 1984 Conference on Teleology in Natural Science* (Center for Philosophy of Science, Pittsburgh, 1985).

17. M. Gardner, WAP, SAP, PAP, and FAP, *New York Review of Books*, **33, 8**, 22 (1986).

18. H. Reeves, The Growth of Complexities in the Expanding Universe, in *The Anthropic Principle*, Proc. 2nd Venice Conf. on Cosmology and Philosophy (Nov. 1988) ed. F. Bertola, U. Curi (Cambridge University Press, 1993).

19. T.J. Kuhn, *The Structure of Scientific Revolution* (Chicago University Press, 1972).

20. P. Duhem, *Aim and Structure of Physical Theory* (Princeton University Press, 1954).

21. D.V. Lindley, *Use of Prior Probability Distributions in Statistical Inference and Decisions*, Proc. 4th Berkeley Symposium on Mathematical Statistics and Probability, **1**, 453 (California University Press, Los Angeles).

22. E.P. Wigner, The Unreasonable Effectiveness of Mathematics in the Natural Sciences, *Comm., Pure and Appl. Math.*, **13**, 1 (1960).

23. P.A.M. Dirac, The Evolution of the Physicist's Picture of Nature, *Scientific American,* **208**, No. 5, 45 (1963).

24. S.W. Hawking, *Phys. Rev.*, **D 37**, 904 (1988).

25. B. Carter, *The Anthropic Principle: Self-selection as an Adjunct to Natural Selection in Cosmic Perspectives*, ed. C.V. Vishveshwara (Cambridge University Press, 1988).

26. P.C.W. Davies, *The Accidental Universe* (Cambridge University Press, 1982).

27. D.W. Sciama, The Anthropic Principle and the Non-uniqueness of the Universe, in *The Anthropic Principle*, Proc. 2nd Venice Conf. on Cosmology and

Philosophy (Nov. 1988) ed. F. Bertola, U. Curi (Cambridge University Press, 1993).

28. J. Leslie, Cosmology, Probability, and the Need to Explain Life, in *Scientific Explanation and Understanding*, ed. N. Rescher (University Press of America, Lanham and London, 1983).

29. J. Heidmann, The Anthropic Principle and the SETI Perspective, in *The Anthropic Principle*, Proc. 2nd Venice Conf. on Cosmology and Philosophy (Nov. 1988) ed. F. Bertola, U. Curi (Cambridge University Press, 1993).

The Growth of Complexity in an Expanding Universe

HUBERT REEVES

Section d'Astrophysique CENS – F91191 Saclay
Institut d'Astrophysique de Paris, 98 bis Blvd Arago – F75014 Paris

1. A Principle of Complexity

The subject-matter usually covered under the title 'The Anthropic Principle' does not, in my opinion, justify its denomination. The 'initial conditions' required for the advent of our species in the universe are also required for the advent of the dolphins or the black-capped chicadees. In fact the mere existence of a sugar molecule sets almost the same constraints on the so-called 'coincidences': the numerical parameters describing both the global properties of the hot primordial soup and the constants of the laws of physics governing this matter.

I prefer the expression 'A Principle of Complexity'. If they could speak, the numberless species of plants and animals sharing our terrestrial habitat may also prefer this less chauvinistic point of view. Here however I will keep the traditional name, abbreviated into the A.P.

2. A Predictive Principle?

I plan to discuss an interesting application of the A.P. in relation to the growth of complexity in an expanding universe and with the second law of thermodynamics. I will try to show that the mere existence of complex systems in our world restricts the cooling rate of the universe to be within certain bounds.

If the cooling is too **slow**, the universe remains in a state of reaction equilibrium with respect to nuclear, weak and electromagnetic processes, precluding the formation of non-equilibrium structures on the small scale: nuclei, atoms, molecules and living organisms. Matter would be composed of iron.

If the cooling is too **fast**, the universe remains at all time in its initial state of maximum gravitational entropy, precluding the formation of non-equilibrium structures on the large scale: galaxies, stars, planets. The absence of stars would then also preclude the formation of nuclei, atoms, molecules and living organisms.

The rate of cooling of the universe can be derived from Einstein's theory of general relativity. The rate is, as expected *a posteriori* from the existence of complex structures, just between the two limits quoted above.

I take this example to illustrate what could be the right use of the Anthropic Principle. It has often been said that the principle is not predictive. I do not agree. The previous example proves the opposite. Predictions can be made from *a posteriori* considerations. Our existence, just as the existence of dolphins or sugar molecules, requires the existence of stars, and also of non-equilibrium nuclear and electromagnetic reactions. Therefore it *predicts* that the rate of cooling must be between the numerical limits to be discussed later.

The wrong use of the Anthropic Principle would be to stop the investigations at that point, considering that we have, by the anthropic reasoning, a sufficient explanation of the behaviour of the cooling rate. The right attitude is to try and derive this predicted property from theoretical physics. In the case discussed here this was already done before the discovery of the expansion of the universe.

Another example comes from an historical *a posteriori* prediction by Fred Hoyle. From the fact that carbon and oxygen are both abundant (almost equally) in the universe, he made specific predictions about the properties of the excited levels of ^{12}C and ^{16}O. These levels were later studied experimentally, and found to have the foreseen properties. Again, it would have been most unsatisfactory to consider that the similar abundances of carbon and oxygen were sufficient proofs of the existence and properties of these levels.

This brings us to another aspect of our subject which is best introduced by the following remark of Weinberg: 'I certainly would not give up attempts to make the anthropic principle unnecessary by finding a theoretical basis for the value of all constants.' Behind this statement stands the dream of all physicists: the formulation of an 'ultimate theory' of matter, from which the laws of physics, their associated coupling constants together with the so-called initial conditions of the universe could be derived unambiguously. At the present time it is fair to admit that we are far away from such a formulation.

In our present state of ignorance, we must face alternatively two possibilities:

(1) Such a constraining formalism may not possibly exist and the physical parameters describing our universe are only one set out of many possible sets. This hypothesis is implicit in the popular many-universe interpretation of the A.P. (occasionally called the strong A.P.). We just happen to be in a universe which has a 'fertile set' of numerical parameters, i.e. a set which allows the growth of complexity. My own view on this interpretation is that, unless there is some way, no matter how indirect, to test this hypothesis experimentally, we are in the domain of science fiction. In this

spirit of Ockham, I would say that this is a very uneconomical way of solving a nagging problem.

(2) The second hypothesis: an ultimate theory, which just waits to be formulated by astute physicists, does exist and implies that our 'numbers' were not chosen at random. They are the only possible set compatible with this ultimate theory, thus rendering the many-universe interpretation useless.

So we are faced with the following situation: either the cosmological numbers are chosen 'at random' and we happen to have obtained a fertile set, or they can be derived from an ultimate theory which constrains them to take their values and these constrained values are *just right* to promote the growth of complexity. Personally I find both possibilities *equally astonishing and perplexing*.

3. Heat Death or Heat Limbos?

In 1865 the German physicist Clausius coined the term 'heat death' to describe what appeared to be the inevitable fate of the universe. Following the recently discovered Second Principle of Thermodynamics, all the observed temperature gradients in the world were expected to decrease with time. Consequently matter would eventually reach a state of isothermy, which was believed to be the state of maximum entropy. All particle reactions would then be in equilibrium, precluding complexity and all life.

To the question: 'if this is the inevitable fate of the universe, why is it not in that state?' the only possible answer would have been: 'because the universe was, in the past, in a state of even stronger disequilibrium'.

Just a century later, in 1965, the discovery of the fossil radiation by Pensias and Wilson under the form of a 3 K background, turned the situation upside down. The state of isothermy was seen to lie not in the future but in the distant past. The pertinent question became: how did the universe emerge from these primordial 'thermal limbos', to give rise to the present state of strong thermal disequilibrium, appropriate for the growth of complexity?

In order to follow this interrogation, it is useful to introduce the concept of 'information' à la Shannon. For a given closed system, information will be defined as the (negative) difference between its actual entropy and the maximum entropy it could possibly take:

$$I = S\,(max) - S\,(actual) \tag{1}$$

One fundamental discovery of physics since Clausius is the role of gravitation in relation to the second principle of thermodynamics. Consider a gas of N free particles with total mass M and temperature T enclosed in a volume of radius R. The state of isothermy is the maximum entropy state of this system *if and only if* its thermal energy NkT is larger than its self-gravitational energy GM^2/R. Otherwise this mass is unstable toward collapse. This phenomenon, called the 'gravithermal catastrophe' after Linden-Bell, generates ever increasing thermal gradients. It is the source of all thermal gradients in our universe.

As time goes on, the collapsing structure becomes more and more condensed. If it is massive enough (more than a few solar masses), it may collapse all the way to the state of a black hole.

The thermodynamics of this situation is best understood by introducing the concept of gravitational entropy. The total entropy is then the sum of the (ordinary) thermal entropy and the gravitational entropy.

Gravitational entropy is not easily computed. In fact, in most cases, the computations are still plagued with many difficulties. However, thanks to the work of Hawking and Bekenstein we can evaluate the maximum entropy state of a mass M that corresponds to a black hole of that mass.

First a temperature is associated with the black hole in analogy with a standard thermodynamical 'black body', stating that photons emitted by such a body have typical energies $E = h\nu = hc/\lambda \approx kT_{bh}$. Considering that the radius R_{bh} of the black hole is the smallest distance on which we may have any information, we may infer that the wavelength emitted is of the same order. Since the radius of the black hole is proportional to its mass

$$R_{bh} = 2GM_{bh} / c^2 = 2M_{bh} / M_{pl}^2 \tag{2}$$

we have

$$kT_{bh} = hc / \lambda \approx hc^3 / GM_{bh} = M_{pl}^2 / M_{bh} \tag{3}$$

The Planck mass M_{pl} is about 20 micrograms, corresponding to 10^{19} GeV, or to a temperature T_{pl} of 10^{32} K. The Planck length $R_{pl} \approx 10^{-33}$ cm. For a black hole of one solar mass $M_{\odot}$ we have in kelvin):

$$T \approx 10^{-7} \, (M_{\odot}/M_{bh}) \tag{4}$$

The entropy of the black hole is then given by:

$$S = M/kT = k \, (M_{bh}/M_{pl})^2 \, S/k \approx 10^{77} \, (M_{bh}/M_{\odot})^2 \tag{5}$$

Note that, as in the case of the gravitational energy, and for analogous reasons, the gravitational entropy is proportional to the *square* of the mass.

With these definitions, we can define the information associated with the matter contained in a volume V as the difference between the maximum entropy of this system, and its actual entropy. S_{max} would be obtained if all the mass-energy in this volume could be converted into a black hole: $S_{max} \propto$

$(\rho(T)V)^2$ while the actual entropy is given by: $S_{actual} = s(T)V$, where $s(T)$ is the actual entropy density. The ratio S_{max} / S_{actual} is proportional to $\rho^2 V/s$. In a radiation dominated universe, we have $\rho(T) \propto T^4$ and $s(T) \propto T^3$.

At time t of the expansion of the universe, the correct calibration volume is the 'causal volume' of radius $= ct$, as no information can be obtained from outside of this volume. Thus : $S_{max} \propto T^8 t^6$ and $S_{actual} \propto T^3 t^3$ in the radiation era.

For a volume of Planck radius (10^{-33} cm) at the Planck temperature (10^{32} K), the two components are equal. Indeed the amount of matter contained in such a volume at this temperature corresponds to one black hole of Planck energy (10^{19} GeV), which evaporates in a Planck time (10^{-43} s) to become one photon of the same energy. Thus at the hypothetical earliest time of our universe, where these conditions were met, the cosmic information was zero.

In the Standard Big Bang, the temperature decreases as $t \propto T^{-2}$ in the radiation phase (and $t \propto T^{-3/2}$ in the matter dominated era). As a consequence we have in the radiation phase:

$$S_{max} \propto k\,(t/t_{pl})^2 \tag{6}$$

Let us assume that this expression can be extrapolated all the way back to the Planck time (most likely an unreasonable assumption). One then finds that the maximum entropy is approximately one (one unit of k) at the Planck time and grows as the square of the elapsed time. Remember again that at the Planck time, the horizon contains an amount of energy equal to the Planck mass, which evaporates in one Planck unit of time (for this reason, this object is often called an 'instanton'). What is the entropy of a causal region which contains only radiation (no black hole)? It is given by the number of photons (proportional to T^3 and hence to $t^{-3/2}$) in the causal volume (proportional to t^3). Numerically we have:

$$T_{rad} = M_{pl}\,(t_{pl}/t)^{1/2} \tag{7}$$

$$S_{act} = k\,(t/t_{pl})^{3/2} \tag{8}$$

$$S_{max}/S_{act} \propto (T^5 t^3_{exp}) \propto t^{1/2} \tag{9}$$

These equations give us a qualitative understanding of the physical situation. As we move towards the Planck time, we have to consider the gravitational effects of the radiation which are expressed by the fact that one photon of Planck energy is indistinguishable from an instanton. In the simple picture presented here, the radiation entropy is comparable to the maximum entropy at the beginning, but *grows more slowly than this maximum value as time proceeds*.

Let us define n as the logarithmic exponent of the cooling rate of a certain quantity f:

$$n = T/f\,(df/dT) \tag{10}$$

In the Standard Big Bang model the exponent of the expansion time is:

$n_{exp} = -2$ in the early radiation universe; $n_{exp} = -3/2$ in the contemporary matter phase.

From the behaviour of the entropy ratios $S_{max} / S_{act} \propto (\rho^2/s) (t^3_{exp}) \propto T^5 t^3_{exp}$ in the radiation era. Given that this ratio is initially close to one, the necessary condition for the information to increase is that $n_{exp} < -5/3$.

During the matter dominated phase we have: $S_{max} / S_{ct} \propto (T^5 t^3_{exp})$, thus $n_{exp} < -1$ for information to keep on increasing.

4. Reaction Disequilibrium

Another condition for the growth of complexity is the requirement of out-of-equilibrium particle reactions leading to the build-up of physical systems far away from equilibrium concentrations of their components. This, in turn, requires that the reaction timescales t_{reac} for the establishment of these equilibrium concentrations be much longer than the age of the universe.

For each type of physical interaction, the reaction rates $P_{reac} = 1/t_{reac}$ are proportional: (a) to the strength G_i of the interactions; (b) to the energy-integrated cross-sections of the processes (σv) and hence to a certain function $f(T)$ of the temperature at which they take place; and (c) to the density of partners N with which a given particle can undergo a given interaction.

$$P = N(\sigma v) = G_i N(T) f(T) \tag{11}$$

During most of the expansion, the particle-density decreases with the *cube* of the temperature ($N \propto T^3$). When the temperature kT becomes smaller than the mass Mc^2, the appropriate expression becomes $N \propto T^3 e^{-M/T}$.

The reaction cross-sections are proportional to some non-negative power of the energy ($f \propto T^m$). For thermal neutron capture reactions (as the reactions leading to deuteron formation), m is zero. For weak interactions, such as the neutrino captures governing the neutron–proton ratio before the first seconds, $m = 2$. Most of the nuclear reactions involved in nucleosynthesis have their rates dominated by Coulomb factors for which $m = d(\sigma v)/dT$ is large, typically $n > 4$. Thus:

$$n_{reac} = -(3 + m + M/T) < -3 \tag{12}$$

As a result, the interaction-timescales increase faster with decreasing temperature than the expansion-timescale: $n_{exp} = -2$.

In the very early moments, the interaction-timescales are *smaller* than the expansion-timescales. All reactions of the type (A + B → C + D) are then in equilibrium with their inverses (C + D → A + B). All populations of reactants are equilibrium populations. The microphysics is reversible.

Sooner or later, however, the two curves intersect. At lower temperatures, the cosmic temperature falls *too rapidly with time* for the reactions to remain in equilibrium. States of disequilibrium set-in, successively, for each force, each type of reaction. From these states, non-equilibrium populations emerge which will lead to the non-zero chemical potentials and affinities essential for the origin of free energies and organizations in the cosmos.

Following the previous line of reasoning, it appears that the condition for the onset of phases of disequilibrium is that $n_{exp} > -3$.

Thus the A.P. predicts that during the radiation era:

$$-3 < n_{exp.rad} < -5/3 \qquad (13)$$

and for the matter era:

$$-3 < n_{exp.mat} < -1 \qquad (14)$$

Along the wishes of Weinberg, as quoted previously, these exponents can be derived from fundamental physics. In fact they emerge simply from the basic idea of general relativity; the curvature of space-time is proportional to the density ρ of mass-energy. The time-curvature is inversely proportional to the square of the age of the universe t^{-2} (which acts as a 'radius of curvature'). The density of mass-energy is proportional to T^3 in the matter era, yielding $n_{exp.rad} = -3/2$.

Here the two parts of the A.P. programme are fulfilled: prediction from the *a posteriori* observation of complexity of the acceptable range of n_{exp}; confirmation from fundamental physics that the theoretical value lies within this range.

5. Cosmic Entropy and Cosmic Information

In the rest of this chapter I plan to give a more detailed treatment of the phenomena accompanying the growth of complexity in the expanding universe. In particular I shall try to show how quantitative we can be on this subject.

The term 'cosmic entropy' may be defined in various ways. Here we shall consider successively: cosmic entropy per unit volume, per covolume, per nucleon, and per causal volume (i.e. within the horizon).

One of the two fundamental equations of the standard Friedman model of the expanding universe (considered as an hydrodynamic fluid of free particles) states that the cosmic temperature T is inversely proportional to the expansion scale factor R ($T \propto 1/R$). Since the entropy per unit volume is proportional to T^3, the implication is that *the entropy per covolume $S_{cov} \propto R^3 T^3$ is constant.*

In other words, except at certain moments when the hydrodynamical approximation breaks down and microphysics is important, the expansion is a reversible process. The expansion does not *by itself* create entropy. But it creates the conditions by which free energy and entropy can be generated. One important factor has already been described: the particularly low temperature dependence of the expansion rate ($\propto T^2$ or $T^{3/2}$) which generates the reaction disequilibrium states of microphysics.

6. Cosmic Entropy Computation: First Method

In a thermal relativistic gas the entropy is, within an order of magnitude, equal to the number of particles. As a rule of thumb, we shall neglect the numerical coefficient and put them equal. Let us compute the cosmic entropy per covolume, choosing, to be specific, a covolume of one cm³, today. The 3 K fossil radiation gives a contribution of approximately four hundred millimetric photons of one meV or so. According to the Big Bang model, the universe should also contain a thermal radiation of neutrinos with temperature of about 2 K. Their population should be close to four hundred and fifty (three families of neutrinos). These two gases give, by far, the largest contribution to the cosmic entropy: $S_{today} = sV \approx 10^3$.

To this, we should also add the populations of photons of stellar origin. The present energy density of starlight is about one eV per cm³. The integrated momentum distribution is not far from that of a grey atmosphere with mean temperature of one eV. Their density is about one cm⁻³ and this should also be an approximate measure of their entropy. These photons were not present at recombination time (at $Z \approx 10^3$). They represent the integrated sum of the effects of all the stellar processes in the past universe. The entropy increase since recombination is $dS/S \approx 10^{-3}$.

Small as it is, this stellar entropy generation is of major importance since it stands behind the elaboration of all complexity and variety in the universe. (We should also consider, in this respect, the population of radio-photons emitted by radio-galaxies. While their number may be large, their role with respect to free energy generation appears to be rather uninteresting.)

7. Cosmic Entropy Computation: Second Method

Another popular way of discussing cosmic entropy is in *units per nucleon*. This is based on the fact that in our familiar universe, photons are created very easily (usually accompanying entropy generating processes), but the creation of nucleons is practically nonexistent (if we except high energy accelerators). Therefore the ratio of photons to nucleons can be used to track the increase of entropy.

The present ratio of photons to nucleons is $\sim 10^9$. The contribution of stellar photons can be computed from the fact that the ⁴He mass fraction has increased by $\sim 5\%$ from the period of primordial nucleosynthesis ($\sim 24\%$) until today (~ 29%). This represents a total nuclear energy yield of (0.05 × 6.8 MeV/N) ~ 0.34 MeV per nucleon, corresponding to an addition of ~3.4 × 10⁵ photons of stellar light (~ one eV) per nucleon to the cosmic background value quoted above. This calculation confirms the previous statement: the entropy increase due to integrated stellar activity is very small, approximately one part in a thousand.

8. Origin of the Cosmic Entropy

It has been customary, in the past years, to discuss the nature of the hypothetical entropy non-conserving early events which would have been responsible for the large value of the cosmic entropy per nucleon (10^9) before the onset of stellar activity. But it should be recalled here that the notion of 'entropy per nucleon' is only useful in circumstances where nucleons are not freely created, namely at temperature $T < 1$ GeV, smaller than the typical nucleon mass. At higher temperatures, we have $n_{photons} \approx n_{baryons}$.

However since baryons are always created in baryon–antibaryon pairs, one can still set up the compatibility in terms of the *baryonic number*, which, in the astrophysical context, is best defined as:

$$B \equiv \{n_{bar.} - n_{antib.}\} / n_{phot.} \tag{15}$$

Even here we meet problems, since we believe that at very high temperatures (very early on) there did occur large amounts of baryonic number non-conserving reactions which would be responsible for the baryonic asymmetry of our present universe (more on this later). Therefore, the present value of the photon to baryon ratio in our universe, $B^{-1} \approx 10^9$, is in reality the result of many different processes to be described in the following sections. The denominator of the ratio presented in eqn (15) is proportional to the accumulated results of all the entropy producing (photons) processes, while the numerator is a measure of all the baryonic number non-conserving reactions. These factors do not appear to be easily untangled.

This fact removes all meaning to the popular belief that the cosmic entropy value of 10^9 is a measure of the sum of all the entropy non-conserving processes in the early universe. In fact, if the (entropy non-conserving) processes traditionally held responsible for baryosynthesis at the temperature of grand unification (GUT) had not taken place, B^{-1} would be larger, and not smaller as expected from this belief.

9. Free Energy from Baryosynthesis

The most popular solution to the question of the present matter–antimatter asymmetry is the so-called Sakharov process. In the frame of the grand unified theory, there exists a particle X with a mass of around 10^{16} GeV, which can decay either into two quarks qq (with probability r) or into one quark and one lepton ql (with probability $(1-r)$). The antiparticle X^* decays into the antipair q^*q^* with probability r' and into q^*l^* with probability $(1-r')$.

The standard requirement that the theory be invariant under the effect of the operator TCP (corresponding to the product of the operation of time T, charge C and parity P reversal) implies that X and X^* have the same lifetime, hence the same total decay rate. But if the theory contains C and CP violating terms in its Lagrangian, r and r' may be different. Assume that $r > r'$. In this case the decay of an $X X^*$ pair will produce an asymmetric result, namely a surplus of $\varepsilon = (r-r')$ of matter over antimatter.

As the cosmic temperature falls below the GUT temperature ($\approx 10^{16}$ GeV), the X and X^* are no longer generated. They decay in the modes described earlier. However if the particle interaction rates are faster (or at least not appreciably slower) than the expansion rate at these early times ($\approx 10^{-37}$ s), the reactions will remain in equilibrium and no net matter–antimatter asymmetry will result. In the opposite case, the X and X^* will decay out of equilibrium and give rise to an asymmetry proportional to ε.

This process generates a non-zero value of the chemical potential associated with matter and antimatter. It would eventually lead to entropy generation if the nucleons were found to be unstable and to release their full rest mass in photons. It does so, at any rate, by opening the possibility of emitting part of the mass of the surviving nuclei during nuclear reactions.

The lack of appropriate theory for grand unification is the reason why the numerical value of ε cannot yet be calculated. Rather the observations are used to constrain the choice of the theories.

The standard scenario is the following. Before the GUT transition period, the universe is strictly matter–antimatter symmetric, $B = 0$. After this period, it incorporates a surplus of one part in 10^9 in matter.

$$\{n(m) - n(m^*)\}/n(m) \approx 10^{-9} \tag{16}$$

Later on, when the temperature becomes comparable to the nucleon masses, wholesale quark–antiquark annihilation takes place, leaving behind the unpaired quarks to be bound into nucleons. Still later, the same thing happens to the electron–positron pairs.

Thus *both* the existence of C and CP violating terms in the Lagrangian of the fields and the basic fact of the expansion, with its own timescale, are necessary for the birth of nuclear and atomic free energy. The ε term is *created* by the C and CP terms and *preserved* by the relatively fast rate of the universal expansion with respect to the relevant reaction rates.

10. Free Energy and Massive Particles

From the previous section we have learned that the presence of massive particles in the universe is intimately connected with: (1) the non-zero value of ε; (2) the comparative timescales of reactions and expansion leading to disequilibrium decays of the X.

(To be complete, we should also discuss the relation of the masses of the elementary fermions and the energy scale of the Higgs field. In the standard theory, the symmetry breaking processes are responsible for fixing the masses of the fermions. However the theory still contains too many free parameters for a fruitful discussion to be possible at this stage.)

The connection between the existence of free energy and the presence of massive particles is apparent through the fact that the binding of physical systems, whether gravitational, atomic or nuclear, is accomplished through the release of a certain fraction of the masses of the formerly free particles.

In a state of reactive equilibrium, this so-called binding-mass-fraction is emitted as soon as it becomes available, bringing no net entropy to the universe. In states of disequilibrium, on the other hand, its emission is delayed. It then represents the form taken by the free energy. It will become, when later emitted in out-of-equilibrium conditions, a new entropy contribution to the cosmos.

The baryosynthesis chapter is, in a sense, a prototype of this behaviour. The matter–antimatter asymmetry generation processes are those to which we owe the survival of some massive particles and, much later, to the generation of free energy, involving nuclear, electromagnetic and gravitational information. Again notice that both the non-zero value of ε and the characteristic timescale of the universe are involved in this occurrence.

11. Nuclear Information

The role of the nuclear force will be treated in detail, as it will be of general pedagogical value for the rest of our story. It will allow us to use the notion of information extended on a cosmic scale. For this, we use the Sackur–Tetrode entropy formula to extend the approximate rule, 'the number of particles in a thermal system is a good measure of its entropy', to the case of an ideal gas.

Consider a gas of nucleons in thermal equilibrium with a heat bath at temperature T. Consider the transformation of 56 free nucleons into one bound ^{56}Fe nucleus (the nucleus with the highest binding energy per nucleon). The energy release dU is 8.6 MeV per nucleon. In the heat bath, this energy yield is rapidly transformed into photons of energy kT.

We divide the universe into two parts: the 'system' under study and the rest of the world. For the initial state of the system we consider a group of 56 free nucleons (an ideal gas) in a given volume of space. In the final phase (pure iron), the entropy has decreased by a factor of 56. But a number of photons N_γ = dU/kT is added to the entropy of the external world. At temperatures below one MeV the increase in the number of photons overcompensates the decrease in the number of massive particles and the fusion is entropically favoured since it corresponds to a positive entropy increase of the universe (the system plus the rest).

If we consider the production of iron in the context of Big Bang nucleosynthesis, we have to take into account also the effect of the large population of photons with respect to nucleons: $(N_\gamma/N_N \approx 10^9)$. The transition temperature is then closer to 0.2 MeV (Kolb 1986). Later on we shall address the question: why is there any hydrogen left in the universe today? The obvious answer is in terms of the very small reaction rates of the hydrogen–iron transformation in the present cold universe. The duration of the process would be many orders of magnitude longer than the age of the universe.

As discussed before, while free nucleons represented the maximum entropy state at high T, as the universe cooled below one MeV the Fe + γ gas became suddenly the most entropic state. Nuclear entropy became *available*. (The words 'nuclear entropy' mean: the entropy added to the cosmic background by the formation of a nucleus from free nucleons. A more accurate expression would be 'entropy of nuclear origin'.)

At this time, the nuclear force *could* have transmuted everything into iron, thereby increasing the cosmic entropy to its new maximum value. It *could,* but it *did* not. Or, more exactly, it went only part of the way towards this 'goal'. Only 25% of the nucleons were involved in the process called 'primordial nucleosynthesis'. And the end-product was not iron but helium, corresponding to only 6.8 MeV (and not the full 8.6 MeV) per nucleon.

As a result, hydrogen is still present in the universe despite the fact that, entropywise, the iron state is favoured. This fact will serve to introduce quantitatively the notion of 'nuclear information' (defined earlier as the difference between the maximum possible entropy and the present actual entropy). According to this definition, the nuclear information is zero above the temperature of around one MeV, since both the nuclear maximum entropy and the nuclear real entropy are then equal to zero (no *net* addition of photons can come from nuclear reactions). Around one MeV, the maximum entropy rises slowly (approximately as 8.6 MeV/kT per nucleon). The real nuclear entropy however remains zero, as no new nuclei are *actually formed*.

This situation lasts until the temperature of 0.1 MeV, when the first deuterons manage to resist the photodisintegrating effect of the photon gas. Rapidly these deuterons are transformed into helium nuclei, but proceed no further in the mass scale. Furthermore, three-fourths of the nucleons manage to escape these nuclear processes. I want to stress the fact that this 'incompletion' is the essential factor giving rise to nuclear information in the universe.

Today the background radiation is at 3 K (one milli-electron volt per photon). The amount of nuclear information is equivalent to six billion photons for each nucleon. Stars are busily spending this information, thereby increasing the nuclear entropy in the universe. The flow of entropy emerging from these stars can be 'tapped' to organize matter in their vicinity.

Let us come back to the building of nuclear structures. There are several reasons why iron nuclei did not emerge from the primordial nucleosynthesis.

Although the weak force is unable to bind any material structure, it plays indirectly a crucial role in the build-up of nuclear information. At temperatures greater than one hundred GeV, the electromagnetic and weak forces have approximately the same strength. Then comes a crucial event, analogous to a phase transition, by which the weak force becomes progressively weaker, due to the very large mass (≈ 100 GeV) of the particle responsible for carrying this force.

One major effect of the weak force is the transformation of protons into neutrons, and vice versa. On the other hand, neutrons turn out to play a major role in primordial nucleosynthesis. Essentially all the neutrons present in the universe at temperatures at which the deuterons survive the heat (below 0.1 MeV) will manage to capture a proton and be transformed, after a few more reactions, into helium nuclei. Thus the building of nuclear information is directly related to the fact that the population of neutrons is *smaller* than the population of protons at the moment of primordial nucleosynthesis. Nuclear information will then be borne by the protons remaining in excess after the pairing.

There is another important factor in the partial failure of the primordial nucleosynthesis in generating heavy nuclei. To build an iron nucleus, one must first associate one proton with one neutron and generate a deuteron. The new nucleons are progressively added, all the way to 56. But deuterons are very weakly bound and do not survive the cosmic heat until the temperature has fallen to 0.1 MeV. At this time, the universe is far too cold to pursue the growth of nuclei, in view of their increasing Coulomb repulsion. An added difficulty is the nuclear instability of nuclei with mass 5 and 8.

We meet here the 'nucleation' problem of very general occurrence in nature. In order for a state to occur, one must first undergo a sequence of intermediate physical processes. These processes may turn out to be impossible, or very slow, at the crucial moment when this state becomes suddenly entropy-favoured. It is well known that the water of mountain lakes often remains liquid well below the zero degree 'freezing point' if it is pure and if the air temperature falls rapidly.

12. Electromagnetic Information

When the universe reached about three thousand degrees, some five hundred thousand years after the Big Bang in the conventional chronology, protons and electrons recombined to form hydrogen atoms. The binding energy (13.6 eV) was emitted in the form of UV photons rapidly thermalized to the background temperature. Since the process took place in conditions of kinematic equilibrium, all the protons were bound in atoms and the electromagnetic entropy created at this moment (13.6 eV per nucleon) was immediately released, resulting in the generation of no *electromagnetic information*.

Later on, stars started to generate heavy nuclei with much higher electromagnetic (atomic) binding energy per nucleon. The full dressing with electrons of an iron atom, for instance, releases in space about 500 eV of photons per nucleon.

When stellar nucleosynthesis started to take place, the universe was already at least a few hundred million years old and the fossil radiation was far too cold and diluted to interact reversibly with the bound atomic electrons. Binding energies can be emitted in out-of-equilibrium conditions, thereby providing appropriate conditions for irreversible processes such as photosynthesis.

Thus it would be correct to say that electromagnetic information appears when heavy nuclei are ejected in cold interstellar space at the moment of stellar disruption in the form of red giants, planetary nebulae, novae or supernovae. On earth, electromagnetic information is stocked in the ground, in the form of oil or coal, the products of ancient photosynthesis, ready to release their energy and entropy in appropriate conditions as provided by furnaces or carburettors. As in the case of nuclear information, only a small fraction of this electromagnetic information can be usefully released by natural processes.

13. The Chronology of Information

Complexity in the universe is borne by many-body bound systems appearing during phases of non-equilibrium (thermal and kinetic) reactions. The corresponding processes lead to the emission of entropy, usually in the form of photons.

The occurrence of non-equilibrium phases is related to the existence of free energy, which in turn depends upon the existence of gravitational, nuclear and electromagnetic information in the universe. Here I present a chronological account of the various physical factors responsible for the existence of cosmic information and entropy.

(a) Planck epoch

Very little is understood about the physics of matter around the Planck time (10^{-43} s, $T = 10^{19}$ GeV). In view of our ignorance, it has become traditional to use the term 'initial data' for various not-understood properties of our universe (bulk properties, laws of physics), tacitly implying that they are the results of unknown processes occurring in this early period.

The existence of our familiar forces of nature (a major factor in the birth of information and complexity) is ascribed to the operation of symmetry groups whose origin can only, at this stage, be considered as one of these initial data.

Contemporary physics deals with the idea of **superstring theories** with spaces with many extra dimensions, the presently most popular version including six new dimensions. These theories (which are still in their early development) seem to present the best chances for the formulation of a coherent unified theory of matter, incorporating the existence and properties of the forces.

Such a theory should fix the properties of all the quantum fields, their vacuum energy states and their associated potential energy terms. These terms, acting as 'cosmological constants' in the expansion equation, give rise, at later instants, to chapters of inflationary expansion generating enormous amounts of entropy. The compactification of the extra dimensions would inject in our 3-D world vast amounts of thermal photons thereby contributing, perhaps importantly, to the present cosmic entropy.

One important characteristic of this early phase is the absence of large amounts of black holes, or in other words, the very low value of its gravitational entropy state. This fact, which is for us more 'initial data', is of major importance for the birth of gravitational information and, indirectly, for all free energies (inflationary phases may also be responsible for the low black hole density).

(b) Grand unification epoch

It was usually accepted, up to a few years ago, that at a temperature of about 10^{16} GeV, the nuclear, electromagnetic and weak forces are 'unified' (one coupling constant). Recent developments, in particular the rise of supersymmetric theories, tend to indicate that the unification energy may be somewhat larger, up to 10^{17} GeV but most likely below the Planck energy (in the framework of the superstring theories the situation may be completely changed).

The processes bringing about the differentiation of the forces (through the fact that the scalar fields find access to their minimum energy states) are responsible for fixing the masses of the fermions. This event is of importance in view of the fact that the transformation of a part of this mass into binding energy by photon emission is the main source of new entropy today.

Before this period, matter and antimatter were equally represented, leading to zero chemical potential for all particles. Baryosynthesis, that is the occurrence of phenomena leading to a small excess of matter over antimatter, is believed to have taken place at the GUT period (more precisely at the reheating period following the corresponding inflationary phase) thereby resulting in non-zero chemical potentials. Without this event, essentially all the quarks would have annihilated later on, giving rise to no baryonic matter excess to display bound state energy spectra associated with nuclear and electromagnetic forces.

(In fact some heavy particles would have been preserved, as the annihilation processes would have gone out of equilibrium at low temperature, but their number would have been far too small for galactic and stellar formation to have been initiated later on.)

As time went on, the coupling constants of the three forces evolved differently. The nuclear coupling $^\alpha N$ rose progressively from a value of about 0.02 to about one, while the electromagnetic coupling $^\alpha EM$ decreased from the same initial value to 1/137. The different behaviour of these two coupling constants is responsible for the birth of these bound energy spectra.

(c) Electroweak epoch

The temperature at which the weak and electromagnetic forces differentiate (about 100 GeV) is related to the value of the masses of the intermediate W and Z bosons. This scale also determines the period of the weak interaction universal decoupling T_F (through the value of the Fermi coupling constant G_F). To preserve hydrogen for later use, it is imperative that the equilibrium conditions last long enough to favour noticeably the population of protons over the neutrons (as otherwise no proton will survive the nuclear burning phase into helium). In other words, the decoupling temperature T_F must not be appreciably larger than the neutron–proton mass difference. An increase of only one order of magnitude in the mass of the W would be sufficient to bring about the exhaustion of hydrogen (unless the n–p mass difference happens to be similarly augmented), leading to a simultaneous emission of practically all the nuclear entropy and hence to no build-up of nuclear information.

(d) Epoch of primordial nucleosynthesis

Many events occur around the first minutes. First, at $T \approx 1$ MeV, the decoupling of the neutrino W (weak) interaction. Second, at $T \approx 0.5$ MeV, the wholesale annihilation of electrons and positrons (except the small surplus of unpaired electrons from the GUT period). Third, at $T \approx 0.1$ MeV, the effective capture of the free neutrons by protons to form deuterons and helium nuclei.

The crucial point, with respect to the birth of nuclear information and free energy, is the fact that the equilibrium weak interaction regime is kept long enough to reduce the relative proton–neutron ratio well below unity, ensuring that most of the protons escape nuclear capture at this time and remain available for nuclear burning in stars.

(e) Epoch of recombination

Two major events take place when the temperature reaches one eV or so, after a few tens of thousands of years. First, the end of the radiation era. Second, the capture of electrons by protons to form hydrogen atoms.

The crucial consequence for the birth of information is the onset of thermal gradients. Freed from the homogenizing effects of the radiation pressure, local condensations of matter can yield to the gravitational pull, thereby emitting gravitational entropy and reaching the high temperatures needed for the onset of nuclear reactions and the emission of nuclear entropy. The lowering of the cosmic temperature, brought about by the expansion, ensures that these entropy-emitting processes will take place in conditions far from thermal and kinetic equilibrium generating thermal and reactional affinities.

This moment is of crucial importance for gravitational entropy, inasmuch as star formation becomes possible but is delayed by the interaction times needed for the actual growth of these massive bodies. Quantum gravitational entropy, stocked since the Planck time, will be released during the evaporation of the black holes.

14. Conclusions

The build-up of complex systems during the lifetime of the universe is associated with quite a number of features which have been discussed by many authors. I have focussed my presentation on the thermodynamic features. Two of them are of particular interest:

(1) The early universe was not in a state of maximum entropy: the thermal entropy was maximum but the gravitational entropy was minimum (very few primordial black holes, if any, perhaps due to inflationary phases).

(2) The rate of expansion was fast enough to allow the continuous growth of gravitational entropy required for the existence of stars, but slow enough to ensure the onset of non-equilibrium regimes with respect to nuclear and electromagnetic interactions, required for the existence of

nuclei (other than iron), atoms (other than rare gases), molecules, cells and organisms.

Whether these features are determined to be so by an ultimate theory of matter, or they just happen to be so in our universe, we do not know.

Bibliography

Barrow J., Proceedings of the Cargese School on *Cosmology and Gravitation*, 1986.

Brillouin L., *Science and Information Theory*, Academic Press, New York, 1962.

Davies P.C.W., *Space and Time in the Modern Universe*, Cambridge University Press, 1977.

Frautshi S., *Science*, 593, **217** (13 Aug. 1982).

Guth A.H., *Phys. Rev. D.*, **23,** 347, 1981.

Hartle J., Proceedings of the Cargese School on *Cosmology and Gravitation*, 1986.

Keith A., Entropy, *Am. J. Phys.*, 52, **6**, June 1984.

Kolb R., Proceedings of the Cargese School on *Cosmology and Gravitation*, 1986.

Prigogine I., *Thermodynamics of Irreversible Processes*, Wiley, 1966.

Schroedinger E., *Statistical Thermodynamics*, Cambridge University Press, 1964.

Sugimoto D., Eriguchi Y. and Hachisu I., *Prog. Theor. Phys. Supp.*, **70**, 1981.

Zurek W.H., *Phys. Rev. Let.*, 1689, **49**, 1982.

The Anthropic and Perfect Cosmological Principles: Similarities and Differences

FRED HOYLE
University College, Cardiff

Let me begin with definitions. By the weak anthropic principle I mean pretty much what has been said already, namely the removal of coincidences and special situations by the circumstances of our own existence. As for instance the similarity between the age of the universe and astrophysical time-scales is demanded by our existence in big-bang cosmology.

For most others, the strong anthropic principle appears to mean that our existence has an active controlling influence on the way the universe is. For me, however, this is medicine I find difficult to swallow. I would rather define the strong anthropic principle as a predictive property. If our existence leads to a potentially falsifiable prediction in the sense of Popper then I take it that the anthropic principle is being employed in its strong mode.

Hubert Reeves has been kind enough to mention my prediction of the existence of a level in the ^{12}C nucleus at about 7.6 MeV above the ground state, and of how this prediction was actually verified. The story turned not just on the need for carbon production but on the need for there to be approximately equal quantities of C and O in the world, making CO a common molecule. Then formaldehyde, H_2CO, becomes an association of the two most common molecules in the universe, whence with only slight rearrangements of atoms one has:

$$(H_2CO)_n \rightarrow \text{sugars and carbohydrates}$$

with $n = 5$ or 6 usually. Thus the crucial metabolic processes of life depended on this issue.

Now let me take a deeper prediction that lies beyond present-day knowledge. The mass of the neutron is slightly greater than the sum of the proton and electron masses. If it were considerably more massive, the neutrons in atomic nuclei would soon decay into protons, which would then fly apart because of electrostatic repulsions – there could be no elements except hydrogen in the world. On the other hand, if the mass of the neutrons were less than the sum of the proton and electron masses, electron capture by protons would cause all stars to become neutron stars, and the Earth would become a neutron planet.

85

The world as we know it therefore turns on a rather precise relation, almost a coincidence, between the neutron and proton masses, a coincidence quite finely tuned. That the triple quark multiplet of 8 baryons should have two members so finely spaced in mass with respect to each other, while the other six members have markedly different masses, is either an almost monstrous accident or the so-called basic coupling constants of physics are not really basic. They would need to be derivable from deeper considerations, with the consequence that they could take values leading to a world very different from that of experience, a different world or worlds which might well be realized elsewhere in the universe, or perhaps realized at different epochs.

The latter is a prediction beyond my knowledge and therefore beyond my powers of calculation, and so not, to my style of thinking, of active interest. The situation is different when one turns to the problem of the biochemical complexity of life. Granted one has available all the basic building blocks of life in essentially unlimited quantities the chances of finding biochemically significant arrangements through assembling the building blocks at random are minute to degrees that we are generally unfamiliar with in the physical sciences. If one is in a generous frame of mind, the chance of assembling a typical enzyme at random might be set at 1 part in 10^{30}. Some 2000 enzymes are needed to operate and replicate even the simplest biological cell, for a chance of random assembly of $(10^{30})^{2000}$. For comparison with this superastronomical number of $10^{60\,000}$ the number of atoms in the visible universe is merely 10^{79} or thereabout.

Such numbers can be coped with cosmologically provided one chooses the right cosmology, namely the steady-state theory. Provided a replicative system can pass between neighbouring galaxies in a time of $\frac{1}{3} H^{-1}$, in the steady-state theory the number of galaxies infected by the system in a time $t \gg H^{-1}$ is e^{3Ht}, which for $t = 10^6 H^{-1} = 10^{16}$ years is $10^{1000\,000}$ galaxies. The replicative system needs to have a size of the order of the wavelength of light. Then radiation pressure from the galaxy of origin provides a repulsive speed from that galaxy of about 1000 km s^{-1}, at which speed the system would travel ten million light years in 10^{17} seconds, the required amount. If the origin of life is contingent on a number of events even so unlikely that $10^{1000\,000}$ galaxies are needed to find any one of them, then they will all become superimposed together in a few times 10^{16} years. Even such exceedingly unlikely organizational events can occur in the steady-state theory. Big-bang cosmology on the other hand has no scope for superimposing one unlikely event on another unlikely event, so that either we are reduced to regarding life as an affair of utterly miniscule chance or we have to postulate the existence of some mystic process that prefers certain complex atomic arrangements to a vast number of other equally complex arrangements, and additionally we have to suppose that the mystic process works *before its end result can be appreciated*. If this teleology be rejected, the evident conclusion is that the big-bang is the wrong cosmology, and that the steady-state is the correct cosmology.

A short proof of this prediction would be the following. The best value of the Hubble constant H appears to be about 90 km s^{-1} Mpc^{-1}, giving $H^{-1} = 11 \times 10^9$ years. Theorists are nowadays insistent on the critical closure big-bang model for which the age of the universe is $\frac{2}{3} H^{-1} = 7$ billion years. But astrophysical ages come persistently in the range 12 to 18 billion years, a clear contradiction. Goodbye big-bang cosmology if one believes 90 km s^{-1} Mpc^{-1} for H.

But what about the microwave background, it will be asked? The case against the steady-state theory, so far as the microwave background was concerned, turned crucially on the assertion that there is no physical particle with adequate emissivity to generate the background astrophysically. This assertion was certainly wrong. Metallic vapours when slowly cooled produce long threadlike particles, so-called whiskers, which if they are long enough have exactly the right emission properties for generating microwaves. Such particles it seems to me also have properties in the far infrared that are suited to providing the opacity source within molecular clouds. When I set up the structural equations for the core region of a molecular cloud, using iron whiskers as the opacity sources, I found two possibilities for the effective temperature of the cloud:

(1) If the particles are long enough for their emissivities to be calculated from the Mie solution for infinite cylinders at all relevant wavelengths, then the effective surface temperature of the cloud must be a little above 3 K.
(2) If the infinite cylinder condition fails for wavelengths longer than $\lambda_0 < 1$ mm, then the effective temperature is given by $T = 4000/\lambda_0$, with λ_0 measured in microns.

The case (1) generates microwaves, while case (2) produces an infrared source.

Subject to these ideas as a conceptual basis, the perfect cosmological principle of the steady-state theory requires the following microwave production rate:

$$E^* = \frac{4H}{\overline{\rho}_{baryons}} a T_{CBR}^4$$

where $\overline{\rho}_{baryons}$ is the average universal baryon density, a is Stefan's constant, H the Hubble constant, and T_{CBR} the effective temperature of the background, according to recent measurements at millimetre and at submillimetre wavelengths (Matsumoto *et al.*, 1988) about 2.8 K. Putting $H = 90$ km s^{-1} Mpc^{-1}, $\overline{\rho}_{baryons} = 3 \times 10^{-31}$ g cm^{-3}, the energy production rate is required to average about 18 erg g^{-1} s^{-1}, or

$$\frac{L}{L_\odot} \cdot \frac{M_\odot}{M} \approx 9$$

very close to Soloman's recent value:

$$\frac{L}{L_\odot} \cdot \frac{M_\odot}{M} = 7$$

for extragalactic molecular clouds. This is a notable predictive success for the above line of reasoning from the strong anthropic principle (in my definition) applied to the immense biochemical complexity of life.

In case (1), where the whiskers have adequate length for the emissivity to be calculated from Mie theory, write T_0 for the effective temperature of the clouds and denote a density factor by α, such that α is the average optical depth of microwaves per unit path length. Two relations between T_0 and α can be obtained:

$$\alpha\left[\left(\frac{T_0}{T_{CBR}}\right)\right]^4 - 1 = 4$$

from the perfect cosmological principle applicable in the steady-state theory and

$$T_0^2 = 0.93\left(\frac{L}{L_\odot} \cdot \frac{M_\odot}{M}\right) \Big/ \left|\frac{d \ln \kappa}{d \ln T}\right|_{T=T_0}^{1/2}$$

where κ is the absorptivity of the particles calculated from Mie theory at the wavelength of maximum emission for temperature T_0. With $LM_\odot/L_\odot M$ taken to have the above value 9 and $|d\ln\kappa /d\ln T| \approx \frac{1}{2}$ from the Mie calculations, the above relations for $T_{CBR} = 2.8$ K give $T = 3.44$ K and $\alpha = 3.13$. The spectrum of the radiation emitted by the clouds is now calculated. Taking account of the slow variation of the absorption coefficient κ through the microwave region one obtains the values given in the following table at the frequencies employed by Matsumoto *et al*. The big-bang values for a microwave black-body temperature of 2.8 K are also given for comparison:

νI_ν *in watts* s^{-1} *ster*$^{-1}$

Frequency (cm^{-1})	8.6	14.1	20.8
Steady-state theory	7.83×10^{-11}	5.17×10^{-11}	1.62×10^{-11}
Observed	7.95×10^{-11}	4.98×10^{-11}	1.82×10^{-11}
Big-bang theory (Black-body 2.8 K)	7.29×10^{-11}	3.35×10^{-11}	5.06×10^{-11}

From this confrontation with observation the steady-state theory wins hand-somely, the opposite to the way popular opinion believes it to be.

Notice the steady-state values are not those of a black-body. This is because the radiation from molecular clouds at greater and greater distances is progressively redshifted. At frequencies less than 8.6 cm^{-1} the two theories are sensibly identical, except at very low frequencies where the Mie calculation must eventually fail, with the consequence that the background must ultimately fall away to values less than a black-body at 2.74 K.

Still thinking of the baryonic material of newly formed galaxies being in clouds, say with masses $\sim 10^3 M_\odot$, within the observable universe there would be $\sim 10^{17}$ clouds. In a narrow observing beam, say of solid angle 10^{-6} steradians, the number of sources would be $\sim 10^{10}$ and the fluctuation from such a beam between two different uncorrelated directions would be $\sim 10^{-5}$.

Here there is an example of a prediction from the strong anthropic principle relating the basic issue of the origin of life to the basic form of cosmology, I am happy to say with a result markedly favourable to the steady-state theory. There seems some slight hope that within the next half-century the religion of the big-bang will die the death it deserves. It will I suspect come to be seen as the biggest impediment to the advance of science to have emerged unhappily in the whole of the twentieth century.

Reference

Matsumoto T., Harakawa R., Matsuo H., Musakami H., Sato S., Longe A.E. and Richards P.L., 1988, *Astr. J.*, **329**, 567.

From the Anthropic Principle to the Subject Principle

MICHEL BITBOL

Institut de Biologie Physico–Chimique
13 Rue Pierre et Marie Curie, 75005 Paris, France

Abstract. The weak anthropic principle allows one to obtain precise values of the cosmological and physical constants, by making a specific use of the fact that the universe is observed by *man*. This outcome has in fact been obtained by establishing a retrospective link between two phenomenal beings: the universe and the human body. Here, it is pointed out that phenomena themselves are appearances for a knowing subject, which cannot be reduced to some aspect of the human body, lest a vicious circle is initiated. A design argument based on this pure subject rather than on man provides further information which is not even tackled by the anthropic principle. In particular, it is shown that it affords an appropriate departure point for generating the overall space-time structure. What we have called the 'weak subject principle' is thus a useful complement to the weak anthropic principle.

1. Introduction to the Subject Principle

The anthropic principle has led one to transform the post-Copernican trend towards relocation of man at a privileged position into an instrument of unification of knowledge.

This principle lies somewhere near the common end of three lines of thought: the first one concerns the methodological analysis of science, and especially the role of measurement devices. The second one is ethical and deals with the position which the scientific view of the world ascribes to man. Finally, the third one is philosophical and raises the following problem: what is the nature of this observer who seems to be more and more involved in scientific thought?

I shall argue hereafter that even though man lies *near* the aforesaid end, he is definitely not identical with it. There, instead of man, we shall find a minimal characterization of the knower, irreducible to man, which could, at this stage, be called 'the transcendental subject' by due reference to Kant.

1.1 The question of the measurement devices

The key argument for the weak anthropic principle consisted in noticing that a scientist should take into account the bias (or selection effect, see Barrow and Tipler 1986) yielded by his instrumentation, before interpreting any phenomenon, and then that the most pervasive instrument a scientist makes use of is his own body, and especially his sense organs.

But one should not stop the progression at this stage. To observe a phenomenon, one needs more than artificial and bodily instruments. It is also necessary to have at one's disposal a set of epistemic regulatory schemes which at least allow one to separate a phenomenon from the shapeless background and to relate it to other phenomena (such as the phenomena by which the *preparation* of an experiment can be checked). From a reductionist point of view, it may be argued that these epistemic regulatory schemes are after all nothing more than the result of phylogenetically or ontogenetically determined axon wirings in the brain, and thus that they point back to man as a biological entity. It must however be noticed that such a reductionism is threatened by circularity, since the existence of the aforementioned 'axon wirings' cannot after all be ascertained without collecting a certain set of phenomena which in turn require the epistemic regulatory schemes. The simplest way to avoid circularity in this case is then to posit some epistemic conditions quite apart from any phenomenal event, and thus to consider seriously the concept of a transcendental subject identified with the set of those conditions.

1.2 The role of man in the universe

The second relevant line of thought begins with the noticing that man occupies a smaller and smaller space-time domain within his own picture of the world. A possible answer to this progressive dismissal of man by himself consists in contending, as Schrödinger did, foreshadowing the anthropic principle, that 'Even if life here be the only life in the universe, then that life is the justification, the focus of the whole thing.'

However, another kind of self-dismissal is even more striking: mind is completely absent from our scientific picture of the world. Accordingly, any other solution of the mind–body problem than 'eliminative materialism' must bring its arguments from outside this picture.

Here again, it was Schrödinger who expressed the most radical reversal with respect to this apparent consequence of the advances in scientific knowledge. For him, mind is not included within the scientific picture of the world because it *is* this picture. In his conception, mind is not only the 'purpose' or 'justification' of the 'whole thing', as life is: it is its very stuff. In particular,

Schrödinger considered that any attempt at locating mind somewhere within the world picture, in a brain for instance, is self-contradictory.

1.3 Who is the observer?

Let us now come to the third line of thought. Some of the most prominent interpretations of quantum mechanics have ascribed a central role to the observer in the very definition of a phenomenon. The difficulty is then to agree about what is called 'an observer'. Here, I can but enumerate some usual positions, in due order, taking as a criterion the level at which the collapse of the wave function is said to be completed or, in other words, the location of the 'Von Neumann boundary'. The order adopted for this enumeration is the distance with respect to the subjective aspects of this observer. The position which, at first sight, is the most extreme one, is solipsism: the only observer is myself. Then comes the more general idea according to which an observer is characterized by consciousness (Wigner). Going a little bit further, Bohm (1951) contended that an observer is essentially a living being with a brain wherein non-linear phenomena take place. Eventually, according to the orthodox Copenhagen position, the observer is, to begin with, a macroscopic system, and, secondly, a communicating being whose language has such limitations that he cannot express clearly the result of any measurement outside the framework of classical mechanics.

In this case, a radical move came from John Wheeler who, illustrating his thought by the description of some 'delayed-choice' experiments, reached the conclusion that the whole universe is *created* by an experiment performed 'here and now', the latter entity having definitely no spatio-temporal coordinate. The replacement of any kind of personal, conscious, living or macroscopic observer by the limiting concept of 'here and now' thus brings back a flavour of participatory anthropic principle. The main difference there is that 'anthropos' is no more the node of the design argument.

We have thus far reached, at the end of the three lines of thought we have followed, a triad of apparently distinct entities. They are respectively: the transcendental subject construed as a set of epistemic regulatory schemes; the mind; and the here-and-now. The three of them are abstracted from some aspect of the role of man. The three of them share an essential feature: they are, by construction, completely *excluded* from the domain they rule (the transcendental subject is excluded from the field of the knowledge it organizes; the mind is excluded from the world which is its representation; and the 'here and now' does not belong to the set of spatio-temporal events it 'creates'). As such, they differ from man who still holds a place, however small, in those domains.

As I shall attempt to make plausible later on, this common negative feature points towards the idea that they may be three aspects of one and the same entity. Let us call it, for short, 'the subject'.

This being granted, at least provisionally, the interesting question to ask is the following: which kind of specific insight, if any, does a design argument based on the 'subject' rather than on man provide? To be worth considering, what I shall call the 'weak subject principle' (WSP) should at least partly fill the gaps left by the 'weak anthropic principle' (WAP). Now, which are these gaps? As is well known, the WAP enables one to determine the value of most (or all) physical and cosmological constants, but it takes for granted the *existence* of the whole space-time framework wherein the phenomena ruled by these constants take place.

Considering seriously Kant's 'Transcendental Aesthetics', the 'subject principle' is then likely to become pivotal in our understanding of time and space.

Before coming to time, in the next section, I would like to give a statement of what I have called the 'weak subject principle':

WSP: *'The topological structure of the world must be such that this world is knowable, and especially objectively knowable.'*

To add that the world must be knowable 'by a subject' would be redundant, since the subject is himself but a set of epistemic conditions for the world to be known.

2. Time and the Subject Principle

2.1 A linguistic outlook

It has long been felt that knowledge cannot encompass *everything*, even though one may be confident that *anything* is *a priori* knowable. The limitation theorems of logics and mathematics have brought this fact within the heart of scientific thought. But, well before the demonstration of these theorems, the aforesaid fact was implicitly rooted in many linguistic habits. A close examination of one such habit may clear up the aspect of the problem we are interested in at the moment.

The splitting between the known (object) and the knowing (subject) is of widespread use in most theories of knowledge. Any attempt at identifying the knowing with some part of the known (a human body, for instance), is bound to be no more than a step in an infinite regress, since even the latter part of the known presupposes a knowing. The aforesaid splitting thus appears to be a very straightforward expression of the fact that not everything can be known, even though there is no criterion to decide that such and such thing cannot be known. Said in another way, a necessary condition for something to be known at all is that there be some unknown range.

The fact that, to indicate this unknown, a *present* tense of the verb 'to know' is used ('knowing') (to be compared with the *past* tense associated with the known) is highly significant. By its own definition, indeed, the content of *now* is unknown, for *as soon as* it is known, it is no more 'now'! Such considerations already suggest that the identification which has been tentatively proposed above, between the (knowing) subject and the (here-and)-now, could be much more than a mere analogy. More arguments are to come.

The basic epistemological requirement: 'that something can be known' may be the departure point for a construction of time (see M. Bitbol 1988a). The aim of such a construction is to overcome the traditional difficulties of both the 'static' view of time and the 'kinematic' view of time. According to the first one, time is but an immutable ordering of events, which is usually expressed by the relations 'earlier than' and 'later than' holding between the events. But, as many authors have rightly noticed, such an ordering does not carry the full content of the concept of time. We are then led to the second conception of time: the 'kinematic' one, according to which time requires 'change' or 'flow'. More precisely, time is supposed to involve the idea of a 'now' *moving* along the immutable series of the events. However, a moving 'now' in turn presupposes a 'supertime' in which its velocity can be defined. Despite many attempts at circumventing them, the paradoxical features of this 'supertime' are still considerable. My own conception, that I have called 'Ekstatic', after Sartre, seems to retain the logical consistency of the 'static' view without losing the irreducible specificity of the experience of change, which the 'kinematic' view tries to preserve. It is based on the considerations which follow.

2.2 The elementary relation of knowledge

As we have just seen, the principle of the incompleteness of knowledge specifies that there must be a boundary between the known and what is not known. But the fundamental fact to be exploited is that it says nothing about the *position* of this boundary. The latter is thus *a priori* arbitrary. Whichever choice is made for this position, it may then be noticed:

(i)   that within the field of the known, there are other possible positions of this boundary, such that they encompass subsets of the content of this field (let us call them 'the inner boundaries')

(ii) that nothing prevents the field of the known from being more extended than the one this choice implies.

As we shall see later on, the first noticing may give rise to the idea of 'succession' while the second one might be the basis of the experienced instability of the present.

To develop the previous remarks in a more formal way, we can write the basic relation $K_G(\ , S_k)$ where S_k is the set of the known events while the blank stands for the 'knowing-unknown'.

Condition (i) them becomes: For any set S_k, there exists a set of m events r_k belonging to S_k, such that the relation K_G also holds between the two following sets: $S_k = \{\ , r_{k1}, ..., r_{km}\}$ and $S_{k-1} = S_k - \{r_{k1}, ..., r_{km}\}$.

Condition (ii) is more subtle. A straightforward interpretation of it is as follows: Saying that the field of the known could be more extended than S_k means that a relation K_G holds between the knowing-unknown and a set $S^u_k = \{\ , S_k\}$ including both S_k and some part of the knowing-unknown:

$$K_G(\ , S^u_k) = K_G(\ , \{\ , S_k\})$$

2.3 Tensed relations

To see clearly the isomorphism between time and the previous relations, which are basic structures of knowledge as a whole, one must further analyse the statements by which existence in time is expressed in ordinary language. These are to be found in the *tensed* sentences.

Simple tensed sentences may easily be expressed through the predicates 'Past, present or future' ascribed to an event e. 'e has happened' ('e happens' or 'e will happen') is accordingly transformed into 'e is past (present or future)'. In formal notations, the latter become: Pe, Ne or Fe (see Prior 1967).

But to express complex tenses (such as past perfect or future perfect), one needs, according to the pioneering analysis of Reichenbach (1947), to consider both the time of utterance of the tensed sentence and the time from which the event spoken of is directly referred to, the latter being called 'point of reference'. More generally, one can account for any tense of higher order than past perfect or future perfect through a hierarchy of points of reference, of which the time of utterance is a particular instance. The construction of the said hierarchy may be carried out in the following way: the time from which the event e spoken of is directly referred to is called the first level point of reference $R_{(1)}$. The time from which $R_{(1)}$'s view of e is referred to is the second-level point of reference $R_{(2)}$ etc ... *As for the time of utterance, it can either remain implicit, in the complex tensed sentence itself, or be explicitly referred to in a meta-sentence. In the latter case it becomes the last-level point of reference $R_{(n)}$ ...*

The introduction of this notion of points of reference may alter considerably the writing of every tensed sentence. Coming back (as the most straight-forward instance) to simple tensed sentences, wherein the only point of refer-ence is the time of utterance itself, two cases are to be considered:

(i) When the last-order (and unique) point of reference is explicitly referred to in a meta-sentence, tensed predicates are easily transformed into *two*-place predicates relating the two components of the couple: (Point of ref-erence, event). For instance, the sentence 'There exists a point of refer-ence $R_{(1)}$ with respect to which e is past' becomes: $\exists R_{(1)}[P(R_{(1)}, e)]$.

(ii) When the last-order point of reference remains implicit, as is the case in most tensed sentences, and especially in simple tensed sentences, a major difficulty arises: how can we speak of a tensed *relation* if one of the terms of the relation is not even referred to? I propose to overcome this diffi-culty in the following way: Saying that an event is past, present or future means after all that it is past, present or future with respect to the *actual now*. But this actual now cannot be said to be simultaneous to any event of the time-series, for this would amount to referring to it in a meta-sentence. To use Brentano's concept of relations, we may call the actual now the 'terminus' of a relation with the event e, which is in turn called the 'funda-mentum' of this relation. In *intentional* relations, the existence of the 'fundamentum' (in the time series, for the case which concerns us) does not imply the existence of the 'terminus' (in the same series). It is enough to say that e has the relational predicate of being past, present or future with respect to an unspecified terminus. In a relational denotation of tensed sentences *which are not part of any meta-sentence specifying he last-order point of reference*, the actual now must then *hold the place of such a point of reference without being specified as such*. To indicate both this place and this absence of specification, we shall denote it by a blank. *Ue* (where U stands for P, N or F) then becomes $U(\ , e)$.

More generally, an nth level complex tensed sentence takes the form:

$$\exists R_{(i)} (i=1, \ldots, n-1)[U^n(\ , R_{(n-1)}) \& U^{n-1}(R_{(n-1)}, R_{(n-2)}) \& \ldots$$
$$\& U^2(R_{(2)}, R_{(1)}) / \& U^1(R_{(1)}, e)]$$

where $R_{(i)}$ are points of reference, & stands for the logical conjunction 'and', and U^i stands for P, N, or F.

For instance, the sentence 'the event e has been future' is:

$$\exists R_{(1)} [P(\ , R_{(1)}) \& F(R_{(1)}, e)]$$

which reads: 'there exists a point of reference $R_{(1)}$ which is past, and which is such that e is future with respect to it'. If the last order point of reference (the time of utterance) of this sentence is referred to in a meta-sentence and called $R_{(2)}$, the sentence becomes:

$$\exists R_{(1)}, R_{(2)} [P(R_{(2)}, R_{(1)}) \ \& \ F(R_{(1)}, e)]$$

2.4 Isomorphism between tensed relations and knowledge relations

Condition (i) of paragraph 2.2 is written as follows: 'For all S_k, the relation K_G holds not only between the knowing-unknown and S_k, but also between the set S_p and the set S_{k-1} defined as in paragraph 2.2.' More formally:

$$\forall S_k, \exists r_{k1}, ..., r_{km} \in S_k [K_G(\ , S_k) \ \& \ K_G(S_p, S_{k-1})] \tag{1}$$

The same proposition being true in turn for all S_{k-1}, S_{k-2}, etc. ..., further steps of the hierarchy might be written. A restriction K of the relation K_G to *single events*, instead of the specified sets S_k, S_{k-1}, S_{k-2}, ...(for instance substituting a single event $r_k \in S_k \cap S_p$ taken as point of reference, both to S_k in the first relation K_G of the expression (1) and to S_p in the second one, and substituting an event $e_{(k-1)} \in S_{k-1}$ to S_{k-1} itself) then yields complete isomorphism with the tensed P-relations. For instance, the restriction $K(\ , e_{k-1})$ of $K_G (\ , S_{k-1})$ corresponds to $P(\ , e_{k-1})$. Moreover, the restriction $\exists r_k [K(\ , r_k) \ \& \ K(r_k, e_{k-1})]$ of the expression (1) corresponds to the complex tensed formula:

$$\exists r_k [P(\ , r_k) \ \& \ P(r_k, e_k)].$$

As for condition (ii) of paragraph 2.2, which was formally written: $K_G(\ , \{\ , S_k\})$, it is also possible to rewrite it as follows. The idea, for this rewriting, is to give an explicit denotation to the unknown part of the set $\{\ , S_k\}$, rather than keeping a mere blank there. We shall suppose that, by analogical projection of the events of the *known* set S_k, this *unknown* part is made of a set of n' events: $e^{u1}_k, e^{u2}_k, ..., e^{un'}_k$. This being granted, $\{\ , S_k\}$ becomes: $\{e^{u1}_k, e^{u2}_k, ..., e^{un'}_k, S_k\}$. A restriction K of K_G to single events may then be defined just as in the case of condition (i). For instance, a possible restriction of $K_G(\ , \{\ , S_k\})$ is $K(\ , e^{ui}_k)$. The (major) difference between the latter and a relation such as $K(\ , e_{k-1})$ is that $K(\ , e^{ui}_k)$ is a relation between the knowing-unknown and an event e^{ui}_k which is itself but an artificial construct aiming at denoting the part of the unknown which was added to S_k to *play the role* of the known in the relation K_G. The isomorphism with a tensed relation F is here obvious, since the latter holds between the 'actual now' and some projected event which does not belong to the set of the known events.

To express this in ordinary words, we shall say that, given a certain boundary between the known and the knowing-unknown, there are two ways of conceiving its alteration: the first one, which consists in transposing part of the known into the field of the unknown, is the basis of the concept of past. The second one, which consists in transposing part of the unknown into the field of the known, generates the concept of future.

Now, even though it is essentially 'static', the previous conception may help to grasp, at least intuitively, the reason for the usual 'kinematic' feeling that time flows: the condition (ii) of paragraph 2.2 says that nothing prevents the field of the known from being more extended than it is actually. This means the possibility to know something of the knowing-unknown. But *as soon* as this something is known, it is *no more* part of the knowing. This self-referential loop in knowledge (whose expression in ordinary language is easier if it involves implicit reference to time through the use of 'as soon' and 'no more') is likely to be the source of the felt instability of the present. Even more metaphorically, I would say that the unavoidable omnipresence of the unknown as a condition for something to be known gives rise to a vertigo. This image of vertigo in turn suggests the feeling of a motion in a *motionless* state, generated by the mere possibility of falling in the adjacent void (the unknown).

3. Objective Events and Space-time

We have just seen how the mere requirement that something be known may generate a concept of time. But the latter is rather close to what is generally referred to as 'subjective time'. A further set of requirements, being conditions for one to construct an *objective* knowledge, does not merely yield a transformation of the previous concept of time: it can be the best departure point for building a complete space-time structure.

Here I can but give a very short sketch of these requirements (Bitbol 1988b). They all aim at organizing the field of the known.

The building bricks of this field are first considered to be the 'events', defined as follows: they are propositions specifying both a necessary condition (or initial condition, including the preparation of an experiment or more generally the context or an observation) and a result (or observation). Then, the idea according to which some sets of these simple events are to be considered as 'local aspects' of a single objective event, is formalized by defining a relation of equivalence called 'similarity' between the simple events, and calling 'objective events' the equivalence classes of simple events by this relation. A relation of posteriority between the objective events is constructed from a relation of inclusion between the corresponding simple events. Eventually, the whole (relativist) space-time structure is generated, following Mehlberg's axiomatics (1980), by defining classes of simultaneity of objective events as those sets of (objective) events which are *not* related to one another by the relation of posteriority.

4. Conclusion

To close this introduction to the concept of 'weak subject principle', I would like to locate it within the main streams of modern thought, by contrasting it with the 'weak anthropic principle'.

The weak anthropic principle derives from:

(i) An ethical trend towards a new kind of centrality of man in his own picture of the world,
(ii) A maturity of biological knowledge which, from now on, relies on an extensive application of the physical laws to the phenomena it studies.

It establishes a retrospective link between two sets of phenomena: the universe and the human body.

On the other hand, the weak subject principle is deeply rooted into recent moves in the philosophy of mind, which were themselves suggested by the advances in cognitive and computer sciences. These moves consist in separating clearly 'Mind' from any kind of one–one correspondence with a material structure, be it a human body, and also from any spatio-temporal location (see Hofstadter and Dennett 1981). Mind thus tends to look more and more like the abstract epistemic subject which is the basis of the WSP: a condition for the existence of phenomena, rather than a phenomenal entity. Being freed from any spatio-temporal links, it becomes logically able to be the actual *foundation* of the space-time structure.

References

Barrow J.D. and Tipler F.J. (1986) *The anthropic cosmological principle*, Oxford University Press.

Bitbol M. (1988a) "The Missing now", (Submitted).

Bitbol M. (1988b) "Events and space-time" (in preparation).

Bohm D. (1951) *Quantum theory,* Prentice-Hall.

Brentano F. (1988) *Philosophical investigations on space, time and the continuum*, Croom Helm.

Hofstadter D.R. and Dennett D.C. (1981) *The mind's I*, Basic Books.

Mehlberg H. (1980) *Time, causality, and the quantum theory*, Reidel.

Prior A (1967) *Past, Present and Future*, Oxford University Press.

Reichenbach H. (1947) *Elements of symbolic logic*, Macmillan.

Schrödinger E. (1931) Interview to the *Observer*, January 1931, In: *Collected Works*, Vienna 1984, vol. 4, p. 334.

Schrödinger E. (1958) *Mind and Matter*, Cambridge University Press.

Wheeler J.A. (1983) 'Law without law' in: Wheeler J.A. and Zurek W.H. eds. *Quantum theory and measurement*, Princeton University Press.

Wigner E. (1961) 'Remarks on the Mind-Body question', reprinted in the volume quoted above.

The Anthropic Principle: A Critical View

LIVIO GRATTON

Professore Emerito
Università di Roma

Abstract In this paper it is argued that the Anthropic Principle does not give a causal explanation of the so-called coincidences concerning the values of many physical constants. Hence the Strong Anthropic Principle is useless from the point of view of the Galileian and Newtonian epistemology.

At other opportunities I have expressed an unfavourable point of view with respect to the Anthropic Principle.[1,2] The reason for my position is that it is not a *physical* principle, although some may think it not proper to call it *metaphysical*. According to Davies[3] it is **anthropic**, whatever this word may mean; in my opinion it is a *solipsistic* way of reasoning, it does not explain anything; it cannot be verified (nor falsified).

I am not a specialist in epistemology. My views concerning (physical) science, which are by no means new nor original, come from my readings of several authors, like Popper, Kuhn, Piaget, ... not to mention Galileo, Newton, Berkeley, Hume, ... although I would feel embarrassed to give more precise quotations. Thus I do not need to take your time talking about them. I hope they will appear quite clear from what I am going to say.

It seems that the first formulation of the Anthropic Principle (which I shall call the Mild Anthropic Principle, MAP) was that any statement concerning Cosmology is biased by a number of circumstances, among which is the present state of scientific development.[4]

Obviously nobody can object to this. Indeed our description of Nature (not only Cosmology) is limited by our present knowledge, both theoretical and observational; it is not absolute nor definitive, but only provisional and transient. Hence if we take the MAP as a warning to cosmologists, the worst we can say against it is that it is unnecessary, since any true scientist must be prepared to change his views whenever he feels compelled by new observations or better theories. Perhaps one should explain what he means by *better* theories, but this is a different matter.

The warning may have a somewhat deeper significance: the bias is due also to the organization of the human way of thinking (that is, our **reason** or Kant's *Vernunft*), at any stage of its evolution.

Later on, however, the Anthropic Principle has been expressed in a different form: the Strong Anthropic Principle (SAP). This is due to the desire to 'explain' a number of **coincidences**, which seem *a priori* quite improbable. They concern the values of the physical constants (Planck's constant, the velocity of light, the constant of gravity, the unit of electric charge, ...), which appear to be too well fine-tuned so that, if one or more of them were changed by one or two orders of magnitude, the whole aspect of the Universe would change; they have been described many times and I do not need to go into details. According to those who support the SAP, they are necessary for the existence of intelligent life in some place and at some time in the Universe.

Even this statement is unobjectionable, if taken literally. Obviously any theory of the Universe must account for its present conditions and we know that the Earth in its present state does support intelligent life; hence (according to SAP) we must be ready to discard any theory from which it would follow that the conditions of temperature, pressure, gravity, chemical composition and so on, which we believe to be necessary for life to exist and which are like those of the surface of Earth, cannot exist at any time in the Universe. But why should we give so much importance to such a rare and so little understood occurrence as life (intelligent or not)? In fact it is sufficient that the Earth with its surface conditions does exist, independent of the existence of life.

The same could be said, for instance, regarding Jupiter; at present it does not seem to be a suitable abode for life and I think it could be very difficult to prove that it will be in the future (even if we knew more about life). Yet it exists and, if from a theory it would follow that a planet having the mass, spin, chemical composition, number of satellites etc. ... as Jupiter has, cannot exist at its distance from the Sun, such a theory could not be accepted, since we know that it does exist. Let me add that it would be much easier to take the existence of Jupiter than that of intelligent life as a test for a theory, since we know much more about the structure and dynamics of massive bodies than about life and its origin and evolution.

This kind of argument holds of course for any existing object. In practice, astronomers find it very difficult to explain the origin of stars and galaxies after the decoupling of matter from radiation, according to the current cosmological theories, and sometimes this is taken as an argument against the *big-bang* by its adversaries. But very few of them, if any, are likely to take the existence of life as an argument in favour of or against their theories.

Apparently the supporters of the Anthropic Principle think that the coincidences, which they believe to be necessary for the existence of life, disclose a kind of **plan**. In other words the structure of the Universe was carefully planned in order to make possible the appearance and existence of **Man** in some place and at some time.

This reminds me of a long wandering passage in Kant's *Universal Natural History and Theory of the Heavens* which the young philosopher seems to be very fond of.[5] '... in the island of Jamaica, as soon as the sun rises so high that it throws the maximum of bearable heat on the earth, shortly after 9 o'clock in the morning, a wind begins to rise from the sea which blows from every direction over the land; its strength increases in the measure in which the elevation of the sun increases. About 1 o'clock in the afternoon when it is naturally hottest, the wind is most violent and subsides again gradually with the setting of the sun, so that towards evening the same stillness rules as at sunrise. Without this desirable arrangement the island would be uninhabitable. All coastal lands which lie in that zone enjoy this same benefit. It is most necessary for them because they, being the lowest regions of dry land, suffer the greatest heat; for the higher regions of the land, where this sea wind does not reach, are less in need of it because their higher site places them into a cooler region of the air. Is not all this most beautiful, are not [here indeed in view] visible goals which are implemented through wisely applied means?'

This excerpt will suffice, although the passage is much longer. Kant's conclusions are that it would be an 'astonishing accident, or rather ... an impossibility' if these useful and goal-directed arrangements '... should have matched one another in their natural tendencies so closely as if a superior wise choice might have coordinated them ... even if no man lived in such an island' (although 'they can be deduced from the most universal and simplest laws of nature').

Is not this an early suggestion of the Strong Anthropic Principle? Kant himself quotes Epicurus as maintaining the same views.

The inconsistency of SAP is that it is not a causal but a *teleological* (or perhaps better a *solipsistic*) argument. Of course everybody is free to accept it, if he likes, but it is just what science has tried to avoid since Galileo's time; I am not able to see how it may advance our knowledge of the physical world.

In fact, after Darwin, every sensible man is aware that plants and animals have evolved in order to adapt themselves to environmental conditions and not, on the contrary, that these conditions have been expressly planned to make their life possible!

It is quite obvious that the philosophical starting point of the SAP is Berkeley's well-known dictum: *esse est percipi* or, as he says '... the very *existence* of an unthinking being consists in *being perceived*'.[6] It follows that the condition for the existence of the whole Universe is the existence of a perceiving (human) mind in some place and at some time. It seems that we are living in such a place and at such a time.

This kind of philosophy, which originates from the seventeenth century's endless discussion concerning the *primary and secondary properties* of matter, is what I call a solipsism.[7] It has been severely criticised by Hume, yet it seems hard to kill. 'There is another sceptical topic of a like nature, derived from the most profound philosophy; which might merit our attention, were it requisite to dive so deep, in order to discover arguments and reasonings, which can so little serve to any serious purpose.' And in a few following lines he shows that in fact the discussion arises from 'the asserting that the ideas of those primary qualities are attained by *abstraction* (Hume's italics), an opinion which, if we examine it accurately, we shall find to be unintelligible, and even absurd'.[8]

In a footnote, Hume goes on: 'This argument is drawn from Dr. Berkeley; and indeed most of the writings of that very ingenious author form the best lessons of scepticism which are to be found either among the ancient or modern philosophers, …; that all his arguments, though otherwise intended, are, in reality, merely sceptical, appears from this *that they admit of no answer and produce no conviction*. Their only effect is to cause that momentary amazement and irresolution and confusion, which is the result of scepticism.'

As a matter of fact, there seems to be an inconsistency between the MAP and the SAP. Truly, we do not know at present a physical explanation of the coincidences which are taken as the motive for the SAP; but would it not be more consistent with the historical trend of the physical sciences to explain not the coincidences themselves, but our marvelling at them, as due to our present ignorance? Who can be sure that in a more or less distant future we will not be able to discover that, after all, they are not improbable and astonishing coincidences, but just necessary consequences of a better (at present unknown) theory?

The coincidences of the centres of the first epicycles of all planets with the Sun, when the first deferents are reduced to the same radius, unexplained by the Ptolemaic theory, led Copernicus to his discovery of the Heliocentric theory and to the birth of modern science. The discovery of the electromagnetic theory of light by Maxwell is a similar instance.

This indeed is a very fruitful methodology, which in the past caused great advancements in science. But I believe that to take refuge in non-physical principles, when we are unable to understand something in physics, is a sceptical philosophy, which should be banished from science. It certainly has nothing to do with Galileo's or Newton's philosophy, nor with Einstein's or Bohr's.

There are two more points which I would like to stress. The first concerns the Kantian admiration for the harmony and beauty of Nature. This is due to a strange, but common, confusion between Nature (or, if you prefer, Reality) and Science, which is only our description of it and is changing continuously. The Psalmist sings: 'The heavens declare the glory of God, and the firmament proclaims His handiwork.'[9] This is certainly beautiful poetry and I cannot find anything wrong in it, but it could hardly be considered as scientific knowledge.

All ancient books (and many modern for that matter), even those written by professed atheists, are full of expressions of man's marvel for the 'order' or 'harmony' of Nature. But Nature is neither 'ordered' nor 'chaotic' (I am not using the words in their thermodynamical sense); only when we try to describe it, we introduce a certain bias (as correctly maintained by the MAP), and if we succeed, we say that Nature is orderly or harmonious.

For this reason the ancients spoke enthusiastically of the 'order of the heavens', meaning the description of the planetary motions through the theory of the epicycles. But when Kant admires the same order, he thinks of an entirely different picture: that given by Newtonian mechanics. An argument often brought forward in favour of Einstein's General Relativity is its aesthetic beauty; this is of course quite true from a mathematician's point of view, but nobody would take it as a proof of its validity as a description of Nature!

It happens also that sometimes it seems impossible to reconcile our aesthetic feeling of order with what experiments do show. Everybody has read Heisenberg's dramatic account of his discussions with Bohr; 'could it be possible for Nature to be as absurd as it appeared in those atomic experiments?'[10] Clearly the fault was not with Nature; but simply the new experiments did not fit the order corresponding to the existing theory, which had itself subsided another different idea of order. When Heisenberg, Dirac and many others created the new Quantum Mechanics a completely new kind of harmony and order was discovered, at the cost of changing some concepts which appeared necessary to Kant's or Laplace's idea of order (the mechanistic conception in this case).[11]

My last words are a warning against a possible **theological** interpretation of the SAP. A rather naive believer might be tempted to take it as an argument in favour of the existence of God, who in His wisdom would have planned the world as a suitable abode for His creature: Man.

In my mind this would be a terrible mistake and almost a blasphemy. Only think of how ridiculous would be the idea of a Creator who first invents General Relativity, the Theory of Fields and something else, perhaps; then adjusts the universal constants exactly to the necessary values for men to exist; and finally, lo and behold, everything starts with a big-bang! Frankly, I would prefer the words of the Genesis, which at least have the great beauty and loftiness of ancient poetry, when mankind was young and the world was fuller of mysteries than today.

Of course a real believer will find in his own heart a much better reason for believing in God; and if he is not able to find it, there is no argument, whether scientific or not, which might convince him.

References

1. L. Gratton, *Cosmologia: La visione scientifica del mondo attraverso i secoli*, Bologna, 1987, pp. 617 ff.
2. L. Gratton, *Kosmos*, Convegno internazionale organizzato dall' Istituto Gramsci Veneto, Venzia, 1987, May 8-9, in press.
3. P.C.W. Davies, *The accidental Universe*, Cambridge University Press, 1982, p. 111.
4. B. Carter, *IAAU Symp.* n. 63, 291, 1974. A recent and concise account of the development of the ideas concerning the Anthropic Principle, with many references, is to be found in: Massimo Fracassini, L. Pasinetti-Fracassini and A.L. Pasinetti, *Astroph. and Sp. Science*, **146**, 321, 1988.
5. I quote from the recent English translation by S.L. Jaki, Scottish Academic Press, 1981, p. 83.
6. G. Berkeley, *The Principles of Human Knowledge*, 1710, end of paragraph 88 (italics are Berkeley's).
7. Solipsism (the view that the self is the only thing that can be known and verified) is usually referred to the single individual; I refer it here to mankind as a whole (or to intelligent life).
8. D. Hume, *An Enquiry concerning Human Understanding*, 1751, paragraph 122.
9. Psalms 16:1.
10. W. Heisenberg, *Fisica e Filosofia*, Ital. transl. by C. Gnoli, Milano, 1961, p. 47.
11. See, for instance, my book, quoted in note 1.

The Anthropic Principle and the Non-Uniqueness of the Universe

D.W. SCIAMA

S.I.S.S.A., Trieste – Italy
Department of Astrophysics, Oxford – UK

Abstract. It is argued that all logically possible universes exist in an ensemble of disjoint universes. An intelligent observer would automatically find himself in a universe whose properties are compatible with his own development. The known fine tuning of these properties would then not imply that such an observer is important in the scheme of things, but simply determines the size of the subset of universes in which he could arise. Although the other universes are disjoint from ours, the proposal that they exist is a meaningful one, because it leads to a specific prediction which could be tested by observations carried out in our universe.

The Anthropic Principle is concerned with the elucidation of man's role in the universe starting from the consideration that, whether or not man is actually important in the scheme of things, it is logically necessary that the universe should have properties which are compatible with his emergence. This consideration has led to a wide variety of different arguments, which have recently been ably expounded by J.D. Barrow and F.J. Tipler in their book *The Anthropic Cosmological Principle* (Oxford University Press 1986). In this article I shall give my own views on these matters, but I emphasise that I make no claim to originality – Barrow and Tipler give full references to both primary and secondary sources.

A good place to start the argument is with the observed coincidence that the present expansion timescale of the universe (for example, the time for galaxies to double their distance from us) is of the same order as the lifetime of a typical main sequence star (that is, the time it takes to burn its chief source of fuel). In the steady state theory there would be no immediate explanation of the coincidence – one would presumably have to argue that the same microphysical parameters which ultimately determine the lifetime of the star are also combined in a similar way to determine the timescale of the universe (which in the steady state theory would be a fixed constant time).

By contrast in the big bang evolutionary theory of the universe there is a trivial explanation, indeed there are two, which I will call the weak and the strong explanations. The weak explanation is based on the fact that the Galaxy still contains vast numbers of main sequence stars which have already formed but which have not burned out. Thus we must be observing the universe at the appropriate epoch after the big bang – about ten billion years – when the expansion timescale would be expected to be also of this order. The strong explanation is based on the fact that **I** exist. This requires me to be near a main sequence star, so it is no surprise that in the present epoch of the universe such stars exist.

This simple but powerful argument leads us to ask the question: **how much** can I deduce about the universe from the fact that I exist? It turns out that various elementary particle, nuclear, atomic and molecular properties of matter have to be very finely tuned for conditions in the universe to have permitted my development – many examples are given by Barrow and Tipler and elsewhere in this book. These finely tuned properties will probably also eventually be accounted for by fundamental theory. But why should fundamental theory *happen* to lead to these properties?

There seem to be three possible answers to this question:

(a) It is a complete chance.
(b) God regards me as such a desirable product of the universe that he has fine-tuned it so as to guarantee my development.
(c) There exist other, disjoint, universes with other laws and constants of nature.

Naturally I must exist in one of the universes whose properties are compatible with my existence, but I have no cosmic significance. This explanation becomes particularly trivial if all logically possible universes actually exist in the ensemble of universes which I am advocating. At first sight this proposal might seem to fall foul of Ockham's Razor. But I believe the opposite to be the case. On the conventional view of a unique universe we have to assume that it was *decided* that all but one of the logically possible universes should not exist. This is a very strong assumption and it is completely obscure how this decision was taken. My own view is that we should invoke as few constraints on reality as is compatible with observation, and that it is this view which is in harmony with Ockham's Razor. Thus I am advocating that everything which is not forbidden is compulsory. My existence, together with the necessary fine-tuning, is then a trivial consequence of this proposal.

I know of three objections which have been raised to this point of view. They are the following:

(a) There could exist many other forms of intelligent life (dolphins, black clouds, etc.). Why concentrate the argument on me (or more modestly, on human beings)?

(b) Theoretical physicists are attempting to explain the precise values of the constants of nature in terms of a fundamental theory. Is not this programme at cross-purposes with the anthropic one?

(c) It is meaningless to invoke other, disjoint universes, since by definition one cannot test this hypothesis by any conceivable experiment or observation.

These objections can be answered as follows:

(a) Dolphins and black clouds might well construct similar arguments. For that matter I could construct such arguments on their behalf. Nevertheless I certainly want to understand why the universe was fine-tuned enough to bring me into existence, and whether I am therefore an important element in the scheme of things.

(b) Each universe would be governed by its own fundamental theory. The one applicable to our universe is certainly worth seeking, but it would not have special significance in the ensemble of universes. There is a possible exception to this statement, which would arise if in fact only one type of universe were logically possible. This would be exceedingly interesting, but would leave us with the anthropic problem unresolved, since it would still require explanation why the one unique possibility is precisely the fine-tuned one.

(c) This objection does not hold because the many universes proposal does lead to a testable prediction. The reason for this is that we would not expect our universe to be a more special member of the ensemble than is needed to guarantee our development. By contrast, a unique universe might be expected to be characterised by very special initial conditions indeed. In that vein Roger Penrose has proposed that initially our universe was conformally flat, and Stephen Hawking has introduced a very special ansatz for calculating the initial condition obeyed by the quantum wave function of the universe. These conditions are mathematically elegant and precise, but do not seem to foreshadow the emergence of intelligent life.

On the other hand, I would expect our universe to be a generic member of the set which could give rise to us. Such a generic universe would not possess a simple mathematical rule governing its initial conditions. The prediction would then be: Penrose is wrong and Hawking is wrong, and this could one day be demonstrable, for example by measuring the initial degree of anisotropy of the universe. A full and rigorous discussion of this question would require a still-to-be-constructed measure theory on the ensemble space of the universes. But I hope it is clear that by making a testable prediction, the hypothesis that there exist many disjoint universes is a physical hypothesis.

The Anthropic Principle and the SETI Perspective

J. HEIDMANN

Observatoire de Paris
5, Place Jules Janssen, 92195 Meudon – France

Abstract After presenting SETI as an observational test for existence of intelligence in the cosmos we define what we mean by intelligence. Then, after presenting Carter's view for the extreme rarity of civilizations comparable to ours derived from his weak anthropic principle, we apply his model to a more favourable case and suggest that delays in our appearance on Earth were due to the slow build-up of the oxygen and of the ozone contents of the biosphere.

We also give a softer outcome to Barrow and Tipler's space-travel argument against the existence of extraterrestrial intelligent life and end by a brief presentation of recent astronomical observations on organic molecules and planetary companions to stars.

We conclude that a temperate balanced evaluation of the situation is a sound attitude and that the SETI enterprise should be pushed ahead.

1. SETI

SETI, the Search for ExtraTerrestrial Intelligence, is an acronym which was coined by Philip Morrison, a professor of physics at the Massachusetts Institute of Technology. He was the first to propose, together with Giuseppe Cocconi, a professor in cosmic rays at Cornell University, in a seminal and now classical paper in *Nature* in 1959, that interstellar communications could be attempted with other eventual civilizations by means of radio waves. Independently, the same year, with the advent of the first large radio-telescope at the Green Bank (Virginia) National Radio Astronomy Observatory, Frank Drake, a young radio-astronomer, started to build a special dedicated receiver to engage in the first search for radio signals from the two closest stars resembling our Sun.

These epochal works were the start of SETI. As a matter of fact SETI is rather the search for information transfer by electromagnetic waves across interstellar distances. This is much more restricted than a search for a grandiose *Intelligence* in the cosmos, so much more grandiose as we know less about it. It is an empirical operational scientific attempt to tackle the vast problem of the eventual existence in the universe of advanced states of evolution having led to results at least as much advanced as what we call *intelligence* in our poor humble human case.

For physical reasons based on the nature of interstellar space and on the nature of electromagnetic waves, and for technical reasons based on our present civilization, SETI starts with the search for radio signals of artificial nature as a first signature of an extraterrestrial intelligence, whether these signals are deliberately sent to us or whether they are just *leaks* from extraterrestrial intelligent activity.

2. Intelligence

After having thus sketched the practical aspect of SETI, this leads us to define what we mean by intelligence. Let me emphasise again a reminder I stressed last year in my contribution at the Cá Dolfin: after 15 billion years of cosmic evolution in hundreds of billions of galaxies each containing hundreds of billions of stars it is impossible to think comfortably that human intelligence is the *ne plus ultra* which the universe could have produced.

In order to carry out our search we have to fight, in particular for money and recognition, with fellow astronomers, scientists, technicians, administrators, senators, politicians, and even philosophers!, and here again we have to adopt a pragmatic behaviour. Then for us intelligence is usually defined by the particular stage reached by life as we know it, on Earth, now.

Essentially it is based on information processing, an information processing which in nature goes from low levels to very sophisticated ones as we go from bacteria, to multicellular organisms to *Homo sapiens* with his intricate central nervous system. However, for us, this nervous system, with its limited kilogram of matter, should not be the *ne plus ultra* and as long as twenty years ago the Soviet radioastronomer Nikolai Kardashev has for this purpose of SETI smashed into pieces the anthropomorphic scale by introducing the concepts of civilisations of types I, II and III. Essentially they are based on the power consumption which they can master: the power available from a planet for type I, from a star for type II and from a galaxy for type III. These vast scales for power are paralleled by comparable vast scales for the masses of matter tractable for information storage and processing. Though we do not know anything, except through pure free speculation, about stages more advanced than ours, these concepts open correspondingly vast perspectives for levels of *intelligence* in the cosmos, far beyond our one kilogram brain. SETI appears for the moment as the only available observational means for us to get information on such eventual fabulously advanced states. Thus, though SETI, down to the daily life of radio-astronomy observatories, is aimed only at the search for radio signals at our intelligence level, it is open to much higher hopes.

3. Carter's Weak Anthropic Principle and SETI

In 1974 Brandon Carter introduced his weak anthropic principle according to which our universe is biased in its observed properties by the fact that we are observers. Here I shall restrict myself only to the 'weak' form, considering the 'strong' form not so appealing; in this I simply follow Carter himself when he wrote in 1983: *Although this (strong) 'principle' has aroused considerable enthusiasm in certain quarters, it is not something that I would be prepared to defend with the same degree of conviction as is deserved by its 'weak' analogue.*

As an astronomer much impressed by the *inflationary universe* cosmologies, like a good number of cosmologists I am much inclined to view the weak anthropic principle in this perspective: the inflationary universe scheme leads quite naturally to the concept of the possible existence of an indefinite collection of different universes. Then the fact that our universe appears to have gone through an extraordinary sequence of happy events which allowed our very existence (Carter's principle): we are in a universe which worked *well* because we are here and otherwise we would not be there to observe it. The situation is quite similar to our situation in our solar system: among the nine planets we are on the only one which happened to behave correctly for us to exist. In this perspective the application of Carter's principle gets indeed a very important observational success, which also gives indirectly some support to the inflationary universe theory.

In his 1983 paper Carter applies the weak anthropic principle to an *a priori* astonishing numerical coincidence: the time it took for *the present stage of emergence of civilization* on Earth is within a factor close to 2 equal to the lifetime of the Sun (4.5 billion and 10 billion years respectively). *Biological processes (...) and astrophysical processes have nothing directly to do with each other,* he notes. And he concludes that the average time intrinsically most likely for the evolution of intelligent observers is much larger than the lifetime of the Sun, with the corollary: *civilizations comparable with our own are likely to be exceedingly rare.* This is very important for SETI, which Carter reduces to an *exciting fiction scenario.*

Here it is my duty to present a more moderate and more balanced view. As a matter of fact, in order to derive estimates, Carter has to introduce simplifying hypotheses to build up a simple stochastic mathematical model giving the probability of appearance of a civilization on a planet. He applies it to the case when 10 critical evolutionary steps are necessary to achieve this appearance and finds a probability 10^{-20}, ultra small indeed. However when I apply it for 3 steps only, and he advocates such small numbers later in his paper from further developments, I find 10^{-5}, quite different! In this case, if there were 1 habitable planet for 300 stars there would exist 10,000 civilizations in our galaxy, worth the effort to search for.

Also, at the start Carter argues that the intrinsically most likely time for evolution of intelligent observers to be very short compared to the lifetime of the Sun is excluded *since it is hard to think of any particular reason why our arrival should have been greatly delayed*. I would like to propose a different view: I have been much impressed by the slowness of the evolution of life after its brisk establishment 3.5 billion years ago; for 2.8 billion years the Earth was populated by monocellular organisms only; the first pluricellular beings, the Ediacarians, appeared a mere 0.7 billion years ago, and the extraordinary so-called Cambrian explosion which led to such a profusion of different life forms started only 0.5 billion years ago. Thus life evolution appears to have been delayed indeed up to 0.7 and 0.5 billion years ago. And here I think of two good reasons for these delays: for the slow build up of the oxygen content in the oceans, which in the end allowed the appearance of high-metabolism pluricellular forms; second the later build up of the ozone protective layer in the atmosphere which allowed the Cambrian explosion on the continents.

4. The Space-Travel Argument

It is my duty also to say a few words about a topic which is only indirectly related to our subject today: the so-called *space travel argument against the existence of extraterrestrial intelligent life*. This is developed at length by Barrow and Tipler in chapter 9 of their book *The Anthropic Cosmological Principle*. This is a learned highly technical work, with over 1500 references, which covers wide aspects of the anthropic principle; it has been greatly publicised. Its impact on a large portion of public opinion has certainly been important. I got some feeling of this even about this conference: when I asked some of my (non-SETI) colleagues whether it would be right to speak on the SETI question, some told me offhand: *there are no ETs, this is shown in Barrow and Tipler's book*.

The essence of their argument is presented at the beginning of chapter 9: *If (extraterrestrial intelligent beings) did exist and possessed the technology for interstellar communication, they would also have developed interstellar travel and thus would already be present in our Solar System. Since they are not here, this implies that they do not exist.*

However, when they develop their argument Barrow and Tipler have to go through a very large number of assumptions, hypotheses or speculations. I counted them along their reasoning and found 112 of them. Now one may ask a very simple question: if each assumption is credible at as much as the 90% level (which I doubt), to which level is the final result credible? For instance if a result rests on 3 assumptions credible at the 90% level each, the final result will be credible at the $0.90 \times 0.90 \times 0.90 = 0.73$ or 73% level. For 112 assumptions the final result is 0.0007% credible, i.e. has less than 1 chance in 100,000 to be true.

Another way to put it is: to which level of confidence must each assumption be credible in order that the final result is just fifty/fifty credible? Answer 99.4%, i.e. each of the 112 assumptions has to have less than one chance in 160 to be false. Even if I were challenged to reduce the number of assumptions from 112 to only say 30, in this case they would not be independent, and each would still have to be 97.7% credible.

This is a level undreamed of in science. The space-travel argument is a most interesting exploratory path indeed; however it does not rule out other paths, in particular the one followed by the SETI endeavour.

5. Conclusion

These last two years SETI has received further encouragement from new results obtained in the bioastronomical branch of research. In particular the analysis of the data collected by the space probe Giotto when it encountered the Halley comet revealed the existence of large amounts of the atoms carbon, hydrogen, oxygen and nitrogen; though the parent molecules of the comet were destroyed by the high velocity of the encounter (70 km/s!), it was somewhat possible to infer the existence of already complex organic compounds such as polyoxymethylene.

Also, evidence for planets around other stars is growing, although the present technology is still not sensitive enough for planets as small as the Earth. Several cases are good candidates for planets 2 to 10 times the mass of Jupiter (which has 300 times the mass of the Earth).

Thus it appears that a temperate balanced evaluation of the situation with respect to SETI is a sound attitude. We should not forget that the question of life elsewhere has deep unclear echoes in the individual and collective subconsciousness of humankind and that we should be very vigilant. Since its start thirty years ago SETI has been much debated, arguments swinging back and forth all the way from no hope to great expectations, staying usually and unfortunately at the speculation level only. However, recently major efforts have been devoted to research and development programmes for new technologies which would be able to yield the most needed observational data which are badly lacking.

In particular the new super-powerful receiver being developed by the NASA SETI Program Office is such a first magnitude technological step; last spring an active proposal from the President of the US was put forward to provide major funding for SETI. In France we have the third largest radiotelescope in the world for the type of radio waves contemplated and I am working hard for a collaboration which we would like to extend to other countries as well.

Would it not be a shame to let pass by the Earth a radio signal which another civilisation might have sent just because we did not listen? And would it not be a pity that scientists did not invite philosophers to ponder on the possibility of other intelligences being discovered in this one cosmos of ours, already so special because of Carter's Anthropic Principle?

I thank with pleasure Brandon Carter for enlightening discussion.

References

Barrow J.D., Tipler F.J., 1986, *The Anthropic Cosmological Principle*, Clarendon Press.

Carter, Brandon, 1983, The Anthropic Principle and its Implications for Biological Evolution, *Phil. Trans. R. Soc. London.* **A 310**, 347.

Heidmann J., 1989, Les Recherches Futures de Signaux Extraterrestres, *Annales de Phys.*, in press.

The Entropic Versus the Anthropic Principle – on the Self-Organization of Life[1]

FRIEDRICH CRAMER

*Max-Planck-Institut für experimentelle Medizin, Abteillung Chemie,
Hermann-Rein-Strasse 3, D-3400 Göttingen, Federal Republic of Germany*

1. The Entropic Problem

Entropy is defined by the second law of thermodynamics. This law is the result of experimental observations about the relation of energy to heat. It describes the fact that in heat transfer or in the transformation of heat into mechanical energy, as in the steam engine, a certain amount of energy is necessarily lost, is dissipated, and cannot be recycled. This is an experimental observation which cannot be subject to doubt.

The difficulty begins when the basic principles of this observation are to be explained. When heat is considered as the mechanical motion of particles, the transfer of heat must be considered as a transfer of kinetic energy in collision. Then, of course, the whole problem is reduced to Newtonian mechanics. The Newtonian apple falls off the tree, down to the ground and eventually further down to the deepest point of the valley. Thus, eventually all things will be at 'sea-level', all matter will have the same temperature, as is defined by the entropic death. However, things are not so simple. In reality, heat cannot be reduced to a problem of Newtonian mechanics. It is impossible to explain the second law of thermodynamics by molecular kinetics (H-theorem). May I recall the almost heroic efforts of the great Ludwig Boltzmann which in the end turned out in failure.[2] Karl Popper remarks with respect to Boltzmann's effort: 'I find Boltzmann's ideas breath-taking in their audacity and their aesthetics. I think however that this idea is entirely untenable at least for a realist. It considers irreversible changes to be illusions. Thereby the catastrophe of Hiroshima becomes an illusion. Our own world becomes an illusion and thus all our efforts to explain our world become an illusion'.[3]

In these few phrases Popper expresses clearly the dichotomy between the entropic principle and the anthropic principle: the automatism of the second law of thermodynamics and the events of Hiroshima occur in different worlds.

Only in a reductionist approach does the entropic death become a reality. This was discussed by Plato 2,400 years ago. In *Phaidon*, Socrates and Kebes discuss 'Life after Death', the cyclic structure of nature and being and becoming as seen below:

Socrates: 'If, in an antithetic couple the becoming of one part of the couple would not correspond to the becoming of the other part of the couple in a typical circular process (*Kreisprozess*), if in contrast there would be only a linear one-dimensional transition, developing from the original situation to the opposite situation without recoupling to the original state, if this would be so, one can easily see that in the end all things would have the same uniform 'Gestalt', a 'non-Gestalt'. There would only exist a completely uniform state (i.e. the entropic death!). All becoming would be terminated. All dynamics come to a halt. '

Kebes: 'How should I understand that?'

Socrates: 'Well, it is not difficult to understand what I mean. If, for example, one would fall asleep as normal, however, there would be no corresponding awakening from sleep, this would demonstrate in the end that parables like the one of the 'Sleeping Beauty' (at Platon 'Endymeon', a male 'sleeping beauty' of Greek mythology) are idle talk and without any meaning, because all other living beings would be in the same situation like the hypothetic non-awakening Sleeping Beauty: All men and things would be in an eventless, eternal sleep. And if all things would be mixed and would not separate themselves again (i.e. the mixing entropy!), if all things would reach the state of maximal dissipation, very soon the word of Anaxagoras would become reality where all things exist at the same time and therefore do not exist at all. And, my dear Kebes, the same would be valid for death and life. If all things which we call living would die, and the dead matter would remain in that state and not re-enter life, then the consequence would be that eventually everything would be dead and no life would exist any more, quite in contrast to our observations.'

2. The Anthropic Problem

The problem is by no means a recent one; it is perhaps a new slogan made rather fashionable through the work of Hawking.[5] In fact it seems to me unfair (and even slightly ignorant) to say that it is recent in the light of humanistic thinking and philosophical tradition since the pre-Socratic philosophers. Protagoras (485–415 B.C.) made the fundamental statement:

panton tôn chrematon metron anthropos

which says:

Man is the measure of all things

Also Descarte's 'Cogito ergo sum' centres the world around the thinking of the human being, although this aspect has been gradually lost in the development of Cartesian science.

If we consider science as such, the anthropic point of view would be the following: Nature is the object of human exploration and knowledge. Terms and measures are set by man. There is no objectively true nature.

On the other hand the entropic point of view would be: Nature has defined the terms and measures. Nature is objective and true. Man is part of this nature. He can recognize only preset facts, connections and laws. In his science he can only solve riddles, to use a term of Wittgenstein.[4] Since however – again according to Wittgenstein – a riddle can always be solved, all scientific problems, everything can be solved in principle. However men, human dignity, the singularity of the individual and the human personality no longer exist. This is the concept of positivism.

The anthropic principle, if I understand it correctly, says that man is more than science, more than the scientific picture of him which can be accomplished now and in the future. We shall consider this more closely later. At the moment we must ask the question:

3. Are the Entropic and the Anthropic Principles Compatible?

In a positivistic (entropic) approach man does not occur. Or better: he is reduced to a behavioural machine which acts according to inherited behaviroural patterns. Ethical values are the result of an optimisation of instincts and may also be inherited. There is no room for the humanistic man, who has been the ideal at least since the Renaissance. I believe that much of the barbarism which this century has shown originates from this positivistic deviation of science, especially from its biologistic variant.

On the other hand, the anthropic approach would make the assumption that *Homo sapiens*, being the final destination of the evolving universe, had been 'planned' from the big bang on and that there is a stringent teleology in the evolution of nature. Nature knew from the beginning that it intended to produce a species which is capable of recognising nature in a conscious manner. Such a primitive teleology is unacceptable to science and is not in accordance with scientific facts in physics and biology. In the end one would have to take refuge in the *Weltgeist*, in Entelechia or in God. The entropic and the anthropic principles seem to be incompatible.

4. Newton and Darwin

The following authentic discussion of Newton has been reported:[6] 'After a discussion about comets Newton tried to convince me that there exists a special kind of rotation of planets. Light and vapours of the sun would clump together and result in secondary planets like the moon, which would grow further and further by attracting more and more matter. Thus they would become finally primary planets and comets which in the end would fall into the sun and fill up again its matter. He was of the opinion that the great comet of the year 1680 after five or six or more revolutions would fall into the sun and as a consequence the heat of the sun would increase so much that all life on earth would be destroyed. Mankind in any case is rather recent, he continued, and there are indications of ruins on earth which indicate such early catastrophes, similar to the one he would predict. Conduitt asked how the earth could have been repopulated after all life on it had been destroyed. "It requires a creator", answered Newton.' Newton believed in a static world, because a creation by a creator is a singular, static act. God after creation stands by and allows the clockwork of the world to run down. In contrast to this is Darwin's view of the world. From his observations he arrived at a completely novel view of nature in that he tries to understand nature as a historical process. By this, Darwin has created a new paradigm in the sense of Thomas Kuhn. What are the consequences? First of all the paradigm is new and does not fit into the previous way of thinking and concepts. This was the technical and economical landscape of the 19th century which in many respects was the Newtonian age, in spite of the fact that Newton had already been dead for more than a hundred years. The consequences of his scientific concept, the technical and machine age had just started to develop and the general *Weltanschauung* of men who used these machines developed accordingly. From the Newtonian concept the modern scientific idea of the world has been formed, the leading ideas of which are causality, mathematical formulation, reversibility, machines and technology. All processes can be repeated experimentally like the vibrations of a pendulum or the trajectory of a stone. Life is a clock which has been wound up in the beginning and ticks out gradually. Unfortunately the process of rewinding has not yet been invented but this surely would come in the near future. Thus the *Maschinendenken* did absorb the idea of evolution and the science of life in general, although prior to Darwin there existed different concepts such as those of Lamarck or Goethe. The Darwinian ideas were conceived as a causalistic scheme from big bang to *Homo sapiens*. This is incorrect and also not consistent with Darwin. The Darwinian theory offers a plausible explanation for the multitude of species and of life. However, it is not a causality scheme. Evolution could have occurred also in different ways. And for the future, Darwinian theory does not offer any prognosis.

5. The Complexity of Living Matter

What consequences does this have for our thinking? Predictability is no longer a criterion for science in such systems as biology, evolution, turbulence, nervous systems, etc. In Newtonian systems, trajectories can be calculated from the starting conditions and can be predicted. Thus, the famous discovery of the planet Neptune in 1846 was celebrated as a triumph and a proof for a materialistic, deterministic view of the world.

The situation has changed fundamentally during the last years, originally starting from quantum physics and relativity theory. We have reached a limit in the description of living structures which can be seen in the analogy to the Heisenberg uncertainty principle in the description of the elemental particles. Evolution is a tree, a system with bifurcation points. This reduces the predictability to certain limited areas.[7] This farewell to scientific prognosis of some physical events which now must be accepted also in the macroscopic world does not, however, mean that science has come to an end. It simply means that one must say farewell to the mythos of prognosis. The Newtonian concept of general linearization of differential equations is an unallowed oversimplification of fundamentally complex systems, which science nowadays tries to attack: life, brain, aging, death. For the description of such systems a new transformation theory is required such as the Baker-transformation. The evolutionary tree can be described only with such a transformation in which unpredictable branching points occur. Similar concepts must be applied to the dynamic function of the central nervous system as a system of hierarchically ordered decision processes with feedback. The first theoretical approaches have been made in this field. And, of course, Eigen's theory of the Hypercycle is such an approach to the problem of evolution.[8] The complexity of life brings us to the limit of our knowledge. Of course we can describe many details of nucleic acids and proteins and their information and structure. However, the cooperation of these components in subsystems and higher organisations consists in an unpredictable network system with feedback for which the characteristics of fundamental complexity are valid.[9]

6. Self-organization

Nobody would doubt nowadays that the universe has evolved, has organized itself and produced all the structures which we encounter. All concepts of evolution of matter and life use the terminus self-organization. Real self-organization is neither laid down in the physical nature of the particles (for example, crystallization), nor is it preorganized through a programme (as an example, the organization of a fertilized egg to the adult organism). Of course no structure can be formed against the physical limits: out of the solution of soap according to the prevailing conditions either a white solid shaving foam or coloured soap bubbles can be formed, but no cotton and no colibri. Real self-organization is a property of a system under particular conditions. A system with a high degree of complexity organizes itself, like the galaxies, the planetary system, the primitive living molecules such as viruses, the higher organisms and the human brain. The self-organization scheme which is relevant to molecular evolution is Eigen's hypercyclic scheme.

What then in reality is self-organization? Where does it exist? At which time and on which substrate does self-organization act? In my opinion, the self-organization of matter finally leading to life can be understood as a physical principle. Self-organization then is a physical attribute of matter since the big bang, in the same way as gravity or inertia is a physical attribute of matter and in the same way as electricity is a physical attribute of electrons. Why and to what purpose matter carries these physical principles and properties cannot, and must not, be answered because it is not a scientific question.

Things are heavy. Therefore Galileo Galilei searched for the corresponding laws of free fall. After they had been found one tried to connect them with other similar empirical laws like the laws of planetary motion and the Newtonian laws, that also are empirical laws. The connecting general theory was then found by Newton with his gravitation theory. Why, however, gravitation as such exists is not a physical question but perhaps philosophical or even a question of belief.

Electricity flows. Therefore one tried to look for the basic unit which flows and found the electron with the negative elementary charge and the mass of 1/2000 of the hydrogen atom. Why, however, an electrical force and field exist is a metaphysical question.

The same applies to the question: why does life exist? Many answers have been given to this question; philosophical answers, religious answers. There is vitalism or dialectic materialism. But all these concepts do not offer scientific answers. All that we can try to do is to find the empirical laws, such as the rules of evolution, the Mendelian laws, the Verhulst law, Eigen's hypercycles, the law of entropy, and then look for a theory which governs these laws. This theory, in my opinion, is: matter has the basic property of self-organization. The capacity of self-organization cannot be separated from matter in the same manner as gravity cannot be separated from matter. The capacity to organize itself does not, however, show up in matter in a uniform manner. This capacity depends on the state in which the particular ensemble of matter is. Living matter exhibits the phenomenon of self-organization the more so the further it is from thermodynamic equilibrium. The same is probably true for all other forms of matter. Matter in this sense is never dead. Living matter by definition is very far from equilibrium or as Prigogine puts it: 'At equilibrium, matter is dull. The further one goes away from equilibrium the more intelligent matter becomes.'

Is this a resurrection of vitalism, which had postulated that living matter in contrast to non-living matter has a *vis vitalis*, which would know what it intends to achieve, which would see its goal before its eyes, namely the accomplished living being or *Homo sapiens* as the crown of evolution? I think self-organization as a basic property of matter is not a revitalisation of vitalism. The theory of self-organization, which I propose here as a basic property of matter in parallel to gravity, is even suitable to overcome the dichotomy between dead and living matter. It makes a pure materialism (like La Mettrie's 'homme machine') obsolete in the same way as it makes the vitalism obsolete. And it is also in accordance with our knowledge in biochemistry, evolution and ontogenesis. However, the present common view about matter must be revised. I shall come to this point later.

7. The Evolutional Field

A general and acceptable understanding of the formation of biological structures requires as a prerequisite the knowledge of material biological structures. This knowledge has grown in the last two decades in an astonishing way. Thus we know most principles although not all details. On the other hand, for the understanding of biological structures one requires the physical and mathematical laws of structure formation which just have currently become more and more evident. 'Formation of form' can probably neither be explained by structure alone nor by mathematics alone. The contrast between materialistic thinking and mathematical thinking is as old as philosophy. Thinking in mathematical terms goes back to Pythagoras and Platon. Are our present theories sufficient for the understanding of the development of life? The one which comes closest to this mechanism is Eigen's hypercycle. In an historical comparison the hypercycle theory corresponds to the Galilean and Newtonian laws of motion, i.e. a mathematical formulation by which all motion (Newton) and all evolution (Eigen) can be described. In order to unify all possible laws of motion Newton created the concept of the gravitational field, which gives an answer to the question: How, when and where is matter heavy? For the organization of life we must now ask the same question: How, when and where does matter evolve? Up to now matter has been completely defined by its mass. The mass becomes measurable in the gravitational field. Newton had postulated this field in order to contract the multitude of mechanical phenomena and give them uniform explanation. With the introduction of this abstract, unreal, even paradoxical concept 'field', immediately phenomena of mass could be explained. 'Field' is defined in physics as the totality of the values of a physical phenomenon which can be attributed to points in space without the presence of a material carrier. Newton had introduced his gravitational field in the year 1686 in order to unify Kepler's celestial mechanics, Galileo's terrestrial mechanics and his laws of motion in a unifying theory. This concept of field which today is very familiar to us was by no means unproblematic when Newton introduced it. Newton, who was highly interested in astrology, in which of course 'action through space' is an allowed concept, had apparently derived his 'field' from there by putting it in exact mathematical form. This was not without contradiction by his contemporaries as, for example, Leibniz who was opposed to the 'magic actions' through space. The concept of field is however nowadays the basis of all theories in physics and as such has not been further interrogated.

For the understanding and unification of living systems I would like to propose the evolutional field in which all events and physical explanations since the big bang, the formation of forms, the evolution of galaxies and men could be inserted. Evolution occurs in three-dimensional space and in time. In evolution irreversible events occur, because of the vectorial character of time and the dissipative structures which are formed. The evolutional field must therefore be at least four-dimensional with time as the fourth dimension. Self-organization is therefore a rather abbreviated expression for a basic property of matter: self-organization in the evolutional field. Self-organization is not a mere 'accident' of matter. It is inseparable from matter and is an attribute of the material substance. Self-organization is the creative potential of evolving matter. Every piece of matter has this capacity depending on its energy level in a dissipative structure. In the following scheme I have summarized this concept of the evolutional field.

8. The Concept of Matter Must be Revised

What is matter? This material substance, which started to evolve at the big bang and has produced all these forms which move around in outer space as stars and on this earth on two legs as men; matter, which can touch upon us and hurt us, which we eat and digest; matter out of which dungheaps and works of art are formed? For 3,000 years people have thought about matter, but never was the concept of matter so reduced and hollow as nowadays. Let us check what spontaneously springs to mind when we hear the word 'matter': hard, heavy, composed of small particles, dull, without mind, available and disposable, transformable through chemistry, transformable through man or natural forces, dead, composed of atoms which again are composed of smaller elementary particles – these in general are the attributes of matter. At no time in human cultural history was the concept of matter so reductionistic. I cannot recall here all the concepts of matter since the pre-Socratic philosophers. One essential tradition in western thinking about matter is derived from Platon. Galileo and even more Kepler refer to Platon when they postulate mathematics as the principle for the explanation of the world. This is even more pronounced in modern physics. Werner Heisenberg had a Platonic concept of matter.[10] This is expressed in symmetry groups and in the *Erhaltungssätzen* of physical parameters. Heisenberg tried to understand the different elementary particles as *Eigenlösungen* of one single non-linear field equation, of which the group theoretical invariants express the mathematical symmetry properties of the elementary particles. This so-called *Weltformel* is considered to be unsuccessful today. However, it throws some light on the concept which modern physics has: the explanation of matter from mathematical principles.

If matter exists in an evolutional field, analogous to the gravitational field, and if matter can exist only in this way, then the traditional concept of matter must be revised. Matter is now conceived as creative, it no longer consists of the inert, hard building blocks of Demokrit, but is receptive to the evolutional field. Matter is non-linear and therefore partially indeterministic, which is also in agreement with quantum mechanics. Matter is pregnant with ideas. At least it can be a vehicle for ideas in the evolutional field. Living matter is always far away from equilibrium. We can simply say living matter and mean by that matter far away from equilibrium which is substantially living. This is not a tautology, because it is a physical property to live. Life is not an 'accident' in the Aristotelian sense nor something stuck upon matter. But it is part of the material substance which appears at that moment when matter is transported far away from thermodynamic equilibrium. Therefore we come back to the concept of matter which originates from Platon and the pre-Socratic philosophers in which there was still no dualism between mind and matter, between the anthropic and the entropic principles.

Natural Laws and Theories in	
Gravitation	**Evolution**
Common Experience	
Matter is heavy and inert	Matter organizes itself, forms patterns
Prescientific Descriptions	
Aristotle: Weight is the number of demokritic atoms	*Aristotle:* Entelechia
	Aquinatus: Self-organization is God'sorganization
Empirical Natural Laws	
Galileo: Free fall, pendulum	Law of entropy, Development of stars
Kepler: Planetary motion	*Mendel:* 1st and 2nd law of inheritance
Newton: Laws of motion	*Eigen:* Hypercycle
Theories	
Newton: Gravitational field	*Cramer:* Evolutional field
Σ: There exists a gravitational field, in which matter is heavy. Gravity (heaviness) or the gravitational field cannot be separated from matter. The gravitational field exists in three-dimensional space.	Σ: There exists an evolutional field, in which matter organizes itself. Self-organization or the evolutional field cannot be separated from matter. The evolutional field has as its fourth dimension the irreversible time.

9. Conclusion

The incompatibility of the entropic and anthropic principles rests upon a too narrow concept of matter, especially of living matter. The entropic principle is prevalent at or near equilibrium. All classical thermodynamics and the first and second laws of thermodynamics refer to situations at and near the equilibrium and therefore deal with dead matter. Modern science now approaching such important problems as life, brain, evolution of the universe etc. has to do with systems far away from equilibrium in which irreversible thermodynamics must be applied. In these systems phenomena of self-organization are observed. In my discussion on self-organization I have shown that with the term 'self-organization' one touches on the metaphysical element of a scientific evolution theory. There are no physics without metaphysical basis, but it is of the utmost importance to define precisely the connecting point between physics and metaphysics in order to avoid a confusion of categories. In the term of evolution the self-organization is this connecting point between theory and metatheory. The result of closer inspection reveals then that the trivial scientific concept of matter must be sacrificed. Why not? In nuclear physics it has been sacrificed for a long time; however, there things are so abstract that they did not become common knowledge until now.

Could God exist in the concepts of science? With my new, broader concept of matter which has been sketched briefly here, I think this question can be answered positively. The biblical story of creation can be neither explained nor denied in an evolutional field theory. In the biblical creation God reveals himself and therefore does not only give an explanation of the world as we scientists try to do, but also he gives a meaning to the world. The question of meaning, however, is excluded in scientific questions and explanations by the premises on which science started.

References

1. For a more detailed discussion see F. Cramer, *Chaos und Ordnung – Die komplexe Struktur des Lebendigen*. DVA, Stuttgart 1988.
2. L. Boltzmann: Über die Unentbehrlichkeit der Atomistik in der Naturwissenschaft. *Ann. d. Physik u. Chemie* **396** (1897), pp. 232–247.
3. K. Popper, *Ausgangspunkte*, Hamburg 1979, p. 233.
4. L. Wittgenstein, *Tractatus Logico-philosophicus* (towards the end).
5. S.W. Hawking: *A Brief History of Time: From the Big Bang to Black Holes*, Bantam Books, New York 1988.
6. The Diary of John Conduitt. R.S. Westfall: *Never at Rest*. Cambridge University Press 1980.
7. I. Prigogine, I. Stengers: *Dialog mit der Natur*, München 1980.

8. M. Eigen, P. Schuster: *The Hypercycle – A Principle of Natural Self-Organization*, Springer, Heidelberg 1978.
9. F. Cramer: Fundamental Complexity. A Concept in Biological Sciences and Beyond. *Interdiscipl. Science Reviews* **4** (1979), pp. 132-39.
10. W. Heisenberg, *Der Teil und das Ganze*, München 1986.

Athropic Biology

MARIO ZATTI

Istituto di Chimica e Microscopia Clinica
Ospedale Policlinico, 37134 – Verona – Italy

1. Faith in Chance

The most common application of the (weak) anthropic principle derives from faith in chance, which faith implies that the Universe is accidental, although improbable. Monod expresses this faith in the following words: 'The biosphere is a strange event, which cannot be deduced from the first principles and is as unpredictable as the particular configuration of atoms which form an ordinary stone'; and also 'proteins are casual structures because if we know the exact order of the 199 residues of a protein, made up of 200, it is impossible to predict the nature of the only unidentified residue. They were created at random by the unwitting hand of chance.'

We must state, however, that there is a difference; if we change the order of the molecules which constitute a stone, it remains a stone. If we change only one amino acid of a protein, its function may be compromised.

For life as we know it, there is a small protein (m.w. 10,000), cytochrome c, which is essential for the entire living world. If we were to wait until the chain of 103 amino acids found in the cytochrome formed by chance in the primitive soup of a terrestrial planet supposing that one attempt took place every second with the 20 different natural amino acids, we would have to wait 20^{103} seconds, i.e. more than 10^{120} years. The improbability is vast for a Universe whose age does not exceed 2×10^{10} years. The alternative is to postulate an infinity of Universes. However, this alternative is irrational, because nothing can be demonstrated by introducing the infinite.

Darwinists, on the other hand, explain that order may be constructed by the process of conservation of favourable mutations. A good example is that of throwing 100 dice, until all of them show the number 4. There are two possibilities:

(1) throwing all the dice every time until we succeed in obtaining 100 4s altogether (in one go);

(2) throwing the first time 100 dice and putting aside all the 4s, and then throwing the remaining dice and again putting aside the 4s until we obtain all 100 dice with number 4.

In order to obtain the desired result with the first way, if we throw the dice once every second, it would take us 6^{100} seconds, i.e. more than 10^{77} years; with the second possibility a sixth of the dice could be put aside after each throw (15 after the first throw, slightly less after the second, about 10 after the third), and it would only take us about an hour to collect all of them. If we give the name 'memory' to the putting aside of events which are favourable by chance, this is exactly what the systems formed by polynucleotides + proteins do with casual molecular fluctuations (mutations). In the same way the 10^{120} years required to obtain cytochrome c could easily be dispensed with if we assume availability of 'memory' and an environment guiding selection.

While this mechanism is acceptable for biological evolution in general, the same cannot be said for biogenesis: in fact, if cytochrome c costs less due to the memory, how much does the memory cost?

Darwinian process means selection of variants with the same probability: we therefore cannot rely on favoured probabilities or on internal causes to explain the emergence of a macromolecule functioning as a memory, having the ability to catalyze its own replication or to make replicas of other similar molecules; such a molecule of nucleic acid (RNA) should have been formed by chance, together with all the other possible random RNA polymers. This is, in fact, what Watson proposes: 'an occasional RNA molecule may have been able to work'.

The first RNA-replicase should have been able to use itself or another RNA as a template for polymerization of the precursors found in the prebiotic soup. Of course, no replicase can copy its own active site, so we must hypothesize that at least two RNA replicases were born contemporarily; and this is plausible because, as Watson observes, it has been demonstrated that as few as 52 nucleotides are sufficient to obtain an accurate catalytic action. Following Watson's theory literally, at least 4^{52} possible different molecules would form by chance, i.e. 2×10^{31} molecules, corresponding to a weight of approximately 5 million quintals; and assuming a productive contact between 2 molecules took place every microsecond, about 10^{18} years would be needed to assure contact between two suitable molecules, i.e. a period 1 billion times longer than the time really available. In addition, nothing assures the permanence of the labile complex eventually produced.

Even if low specificity of the reactions might have made them more probable, the problem of the following interaction with protoproteins, also synthesized by chance, would still remain; such an interaction is necessary, as Eigen showed, because with only one molecular species there would be nothing but a continued series of instabilities. In any case, there must have been the very possibility of interaction, i.e. the dynamic forms, the logic capable of relating the two structural forms, nucleic acids and proteins. Pure chance is not enough to explain reasonably the appearance of the molecular structures of life and their logic. No structure (atomic, molecular, macro-multi-molecular, cellular, etc.) could be formed *durably* without preferential laws, i.e. without intrinsic properties, different from selection and external causes.

Under conditions simulating prebiotic conditions on earth, a variety of fundamentally biological compounds such as adenine, nucleosides, nucleotides, etc. have been synthesized. Murchison's meteorite and also asteroids contain adenine and other organic molecules.

2. Information and Forms

From a theoretical point of view and without specific chemical knowledge, if we wanted to calculate the probability that 5 hydrogen, 5 carbon and 5 nitrogen atoms would unite in the right way to form an adenine molecule, we would obtain such a low percentage as to conclude that adenine formation is practically impossible, unless there is some special organizing force. Glucose has a molecular conformation similar to cyclohexane in which the atoms bound to each carbon occupy the tips of a regular tetrahedron, so that the 6 carbon atoms are not arranged in such a way as to constitute a flat ring, but take on a harmoniously pleated, fairly rigid 'chair' form. Amino acids, when polymerized under favourable conditions, spontaneously form the characteristic α-helix, while polynucleotides arrange themselves in a double spiral, etc. Kenyon and Steinman have spoken about a 'biochemical predestination' of the structural and kinetic forms which gave rise to biological evolution.

Prigogine's fluctuation order, a typical product of evolution of the stationary states of dissipative systems, does indeed substantiate the possibility of reaching new states with reduced entropy, but it is evident that not only are fluctuations amplified in non-linear processes needed, but also states of relative stability which guarantee the permanency of certain structures, i.e. stationary structures, which as such do not have the same probabilities as others, and whose existence is not in fact obvious *a priori*.

Many neodarwinists undervalue the problem of the probability of the initial formation and never take into consideration the problem of the probability of the *permanence* of the structures. On this subject, Webster reminded us about the interior intentionality of Driesch and about the states of organic stability of Yalton and Bateson, while Erwin Laszlo speaks about the necessity of introducing in the theories of development certain types of Universals, almost Platonic forms, conserved as they emerge from evolution in a universal morphophoretic field.

I believe in the existence of a field of forms, beginning with atomic and molecular forms, because the structures possess information which cannot come from nothing. According to Monod, at the base of protein structures there is no information, which he identifies with redundancy when he raises the problem of the unpredictability of the 200th amino acid.

There may be redundancy for two possible reasons:

(1) regular recurrence of symbols;
(2) privileged correlations between symbols.

The difference between a random sequence of symbols and a meaningful text is that while in the former the recurrence of the symbols is casual and therefore equiprobable without any privileged correlation, in the latter certain symbols are more frequent than others and certain letters follow others more frequently: in short, we can see the redundancy. The redundancy is therefore interpreted by Monod as a reduction of disorder (for this reason he affirms that amino acids are distributed in proteins by chance, because there is no redundancy).

In fact, the redundancy decreases the informational entropy of a message. This entropy is reduced to zero in a message formed by the monotonous repetition of only one letter. We cannot say, however, that the latter is a message rich in fantasy and intelligence. It is therefore evident that reduction in entropy may be associated with intelligence, but we cannot say that minimum entropy is associated with maximum intelligence, or that periodicity and symmetry are obvious synonyms of information. Therefore, the identification, theorized by Monod, of the absence of redundancy with chance, is arbitrary: recurrence, symmetry and order may, on the contrary, express limitations of the information content and complexification. For example: redundancy in human language is greater during childhood and decreases later on; symmetry is correlated to the entropy of an isolated system, and maximum symmetry is obtained at equilibrium when entropy is maximal; in a common crystal which is typified by periodicity, the macroscopic structure does not change on moving the atoms in their lattice because it is only a transformation of symmetry, an invariance (of meaning) which presupposes a relatively low information content of the structure.

On the contrary the substitution of only one amino acid in a protein can lead to the destruction of the meaning, or function, due to the high information content which characterizes the structure, as Schrödinger realized when he defined the complex organic molecules as 'aperiodic crystals' (1943). Aperiodic order is the prerequisite for high information content: 'the order is there, but we are not able to express it analytically, and the adjective "aperiodic" is the acknowledgement of our incapacity to express the relation between primary structure and function mathematically' (Luisi).

Even if, after all we have said, we cannot apply the theory of information of Shannon as Monod did, however, even remaining on the same ground, we could formulate an alternative hypothesis in order to understand biogenesis before the intervention of the Darwinian mechanism. At least $\log_2 20 = 4.3$ bits are necessary in order to specify an amino acid at a point of the sequence of a protein. However, in chemical syntheses, we think in terms of calories, but the changeover is easy because the bit, as a measure of the entropy of information, corresponds to $\log_2$ of the uncertainty, while the calorie as a measure of physical entropy is $K_B \log_e$ of the disorder, and therefore it is evident that calories can be used to produce information.

May this not be just what nature does? It is possible that a privileged sequence forms spontaneously, provided the bond energy between the amino acids in this sequence is greater than in other sequences. If nature wanted cytochrome c or a similar protoprotein, with a primary unvariable structure of, for example, 52 amino acids, the information needed would be 224 bits corresponding to $K_B T \log_e 20^{52}$ calories multiplied by the Avogadro number, that is 96 kcal/mol at 25°C (assuming equiprobability of messages). The figure is realistic since the variation of standard free energy for the synthesis of the peptide bond in dilute aqueous solution requires the supply of 2–4 kcal/mol (Fox and Dose). In this context, the necessary standard energy to synthesize a medium peptidic bond of the required specific sequence should be somewhat less (by about the amount corresponding to the acquired information) than that required for other sequences of 52 amino acids, because such a difference would justify the greater probability of the desired sequence. It has been established that by simulating prebiotic conditions, in the presence of non-organic catalysts at a temperature of 60°C, a remarkable repetitive formation of privileged sequences of peptide bonds can be obtained (Fox). RNA chains (rich in A and U), which are relatively unstable, should have been stabilized by joining with proteinoids with specific characteristic sequences. If, then, on a thermodynamic basis, a sort of primitive code was formed and then transcribed, we can hypothesize a logic, a plan by means of conversion of thermal energy to information (this would be impossible if there were 40 amino acids instead of 20). Some fundamental order of the genetic code in the present nucleic acids (for example to codify 50% of cytochrome c) could be the result of a message which started as a privileged sequence of amino acids in the thermal proteinoids, was transmitted to RNA and DNA, conserved there and then returned as a genetic code to form proteins.

It is significant that the experimental (Steinman) formation frequency of synthetic dipeptides which derive from a mixture of amino acids is very near their frequency in the present proteins which are generically codified.

Evolution (cosmic and biological) can be seen as a river which descends following the earth's depressions starting at the top, which is the enormous energy of the big bang, energy used bit by bit to purchase the predisposed forms; the states of relative stability (with emerging properties) are there. The process takes place for example in the stars, where there are the necessary nuclear resonances to synthesize C and O, then in the planets where these elements are used to synthesize amino acids and proteinoids capable of binding them and of forming eventually cytochrome c, etc. Structural information (I_{th}) is gradually increasing throughout the history of the Universe, through atomic, molecular and then supra molecular evolution, while energetic information (I_e) is dissipated. Evolution also means complexification, i.e. increase in variety and aperiodicity, production of configurational entropy, which takes place while the potential electronic energy is transformed into heat. The lines guiding evolution are therefore the 2nd law, together with the use of forms, i.e. states of structural and dynamic stability which nature presents at all levels as predestined forms: the use of the forms is paid for with thermal energy transformed into I_{th}. Thom's 'attraction of archetypes' may be recalled, but postulating the apriority of the forms, from atomic and molecular orbitals to the waves of the sea and to biochemical kinetics, without this meaning the negation of validity of the Darwinian processes whereby information from the environment can be obtained. At the acme of evolution we find the form which constitutes the essence of man, to which Aristotle gave the name νοῦς, the intellect capable of non-discursive thought and of seeing truth.

3. Body and Soul

The psychiatrist, Pierre Debray, writes: 'Since my soul exists, it is something which I believe I know, and I will attempt to say something about it. It is unique, because I do not have a monozygotic twin and because chromosomes which have contributed to the formation of my body contained an absolutely original message in their genetic code. My brain is thus a unique component. But what is more: in this *computer* with its fifteen billion transistors and its umpteen billions of connections, a unique intellectual and emotional personality has formed ...; the assemble of all this on a single tool ... is my soul.'

A computer is without any shadow of doubt an instrument capable of performing operations in terms of *demonstration*, and the problem of mind/brain or spirit/matter can only be solved if we clarify the relationship between demonstration and truth.

As early as 1931, Gödel had demonstrated that for every set of axioms A there exist infinite arithmetical truths not included among its theorems. One of these truths is represented by the statement G which, within A, is demonstrably equivalent to the assertion that 'G is not a theorem of A'. Since the set of axioms is effectively given, the correctness of any proof can be verified by arithmetical methods. Thus, if G *is a theorem of A*, there must be a proof *G* of it such as to be able to prove in A that '*G* is a proof of G', thereby proving in A that 'G is a theorem of A', which means proving the opposite of G in A. Therefore, if A is consistent, G *is not a theorem of A*. But this is what G asserts, so G is true. Consequently, there are true propositions which are not demonstrable.

Now, it is well known that a computer may be constructed in such a way as to operate with certain axioms and formalized rules of logic, thus obtaining any number of demonstrated declarative propositions. We can define these propositions as *true* if we place our trust in the axioms and rules of logic, and we would be tempted to consider the logical sum of the demonstrable propositions as the definition of truth of the computer.

Here, however, we run up against the problem of antinomies, for example the paradox of the liar which has been known since the period of classical antiquity (and at the same time the *a priori* aspect so characteristic of the concept of truth).

Scholars in the field of logic have taught us that in any language we have to distinguish between object language and metalanguage. The word 'truth', and therefore any discussion of it, *must be excluded from object language if we wish to keep it free of antinomy*. Thus, the notion of truth must be *distinct* from the system of demonstrable propositions, and therefore distinct from the computer (and also anterior to it), the computer being regarded as the incorporation of the system of demonstrable propositions.

Even if we speak of consciousness as a property of the neural networks and even if we know the biological processes which have led to, or which correspond to, abstraction, reasoning and language, each of these developments presupposes a notion of truth which is anterior and which cannot be conceived as a property emerging from any of them, or, in other words, from the biological evolution which has constructed the computer. Similar concepts have been proposed by M. Delbrück, though they are not taken up in his lessons on the relationships between mind and matter, published posthumously. What is at issue is the problem of distinguishing between reason and intellect: '... the human intellect, though availing itself of reason to administer its concepts, in conformity with the strictest logic (which derives from its carnal condition), is also an intellect, that is to say a power capable of seeing (as the eye sees) in the order of intelligible phenomena, but with incomparably greater certainty than the eye is capable of in the order of perceptible phenomena. Is it not true to say that it is precisely through such intuition that the intellect knows the 'first principles' of any demonstration?' (Maritain).

St Augustine says: 'Never depart from yourself, return within yourself; truth abides within Man', and he goes on to say: 'We judge by means of truth, but we can never subject truth itself to judgement.'

The soul, which according to Debray is a computer, expresses itself in a language which, indeed, embodies certain elements characteristic of the function of a computer, but which also embodies elements which are not part of such a function and which we are obliged to regard as distinct. In point of fact, the analysis of language and systems of logic (Tarski) shows that the antinomies disappear when, and only when, *the notion of truth is not confused with the operations of reason*. Thus, either we accept antinomies in logic or we must be fully aware, with complete clarity, of what our universal language contains: in this we find the undeniable *expression of a dualism*.

4. Soul and Artificial Intelligence

Some people claim that the paradox of the liar (Epimenide's paradox) owes its absurdity to the fact that it is an operation of self-reference. Gödel's theorem of incompleteness demonstrates that self-reference produces paradox even at the fundamental level of logical analysis. Hofstadter, however, clearly furnishes the demonstration of the reproduction of Epimenide's paradox in the 'Typographical Theory of Numbers', as operated by Tarski, whereby we have a version of the paradox which, at this level, becomes *not an affirmation about itself* but an affirmation about numbers which is true only if it is false.

Self-reference, however, produces paradoxes which can be eliminated if the reference is not made to the *semantic content* of the utterance itself; and this also holds good for the paradox of the liar. Paradoxes depend ultimately on the conflict between two radically different modes used in language to consider the identity of an entity, namely, on the one hand, identity defined by means of spatio-temporal (or logicoformal) localization for a concrete entity, and, on the other hand, identity independent of any related space-time domain and possessing a semantic nature, for an abstract entity such as, for instance, a quality. Entities of the former type are necessarily designated by means of substantives, while those of the second type, corresponding to *qualities* (e.g. truth) may be designated by means of adjectives (e.g. true). This being so, the operation of identification on a semantic basis is alien to the computer and to calculating thought, for which self-reference through the semantic content of a speech utterance is thus prohibited. In other words, qualities, perceived in their semantic identity, are kept distinct (metalanguage) from the language and operations of quantitative logic (object language) and spatial identification, and therefore distinct (dualism) from the subject of action to which the operations in question pertain. This dualism is the reason for the difference which exists between human and artificial intelligence and reflects the difference between truth and demonstration, and, in general, between perception of qualities, essence and forms, on the one hand, and *res extensa*, on the other.

The advocates of artificial intelligence very often neglect the dual aspect (demonstration/truth, reason/intellect) which we have dwelt upon here at some length, identifying all intelligence with the *logic* of the computer (P. Smith-Churchland: 'The mind is the brain'). On the basis of what has been said, this can only be a blind alley, up which people go when they are unwilling to accept the dualistic solution.

When Debray says '... my soul is unique because the chromosomes which have contributed to the formation of my body contained an absolutely original message', what he is saying is right, but incomplete. Aristotle had also established that living things are composed of matter and an information principle (as confirmed by the data of modern biology to which Debray refers) and had reached the conclusion that from this point of view there can be no existence of the 'soul' outside informed matter. Aristotle, however, also recognized the existence of the νοῦς, i.e. of the intellect capable of exercising contemplative cognition, which 'appears to be a different kind of soul ... capable of being separated, like the eternal, from the corruptible'. Moreover, among the various different forms of cognition, Plato had already distinguished the νόησις, or intuitive consciousness. The perception of 'that which is' as compared to 'essence' is therefore proper to the νοῦς, capable of contemplative cognition and corresponding to semantic identification; not being a form of discursive-propositional thought, it is capable of entering into a relationship with the timeless and the eternal.

One absurd consequence of identifying the 'mind' with the computer and of the failure to distinguish between demonstration and truth is that reported by S. Sutherland in a commentary on a book by MacKay. The latter, distinguishing between 'hardware' and 'software' in the brain, suggests that one possible solution to the problem of the resurrection might consist in the incorporation of the programme operating in the brain of each individual in something other than the human body. Furthermore, a hypothesis emerges regarding the possibility of constructing computers which are as intelligent as the individuals of the human species and which, Sutherland observes, could therefore be legitimate candidates for the Kingdom of Heaven.

However, the demonstration, in the form of the mathematical proof furnished by Gödel, that undemonstrable truths exist implies that propositions exist which every human being perceives as being true, for which no human brain can be programmed to provide any logical confirmation: the 'software eye', *pace* MacKay, is not enough to see the Kingdom of Heaven, just as it is not enough to explain freedom.

5. Freedom and Pain

The expression of the evolutionary creativity of the universe is particularly marked in the appearance of biological systems and in the progression leading to the extrinsic manifestation of psychism. At the culminating point of this process, as emerges from the analysis of language, there is something distinct from, and greater than, the mechanics of the computer: there is the possession of truth and, with that, of freedom. *Creative* freedom, for which matter is arranged within the framework of its own measure of elementary indeterminacy; this same atomic indeterminacy, when amplified in the mechanisms of genetic mutation, has expressed the creativity of evolution.

In addition to permitting evolution, mutations are more generally responsible for errors in the formation of genes. Here, 'error' means disease and pain, inasmuch as living organisms are 'the amplified expressions of the molecules of which they are made up'. The structure of living organisms allows us to observe in macrophenomena the effect of atomic 'freedom', of which we therefore have a demonstration. Now, our thesis is this: if Man exists with his freedom, he also necessarily exists with his pain, because both are grounded in the lack of causal necessity governing the possible elementary processes.

On the element of chance or uncertainty involved in the process of becoming, within the framework of systems far removed from a state of equilibrium, interesting studies have been produced by I. Prigogine, who stresses the profound significance of the 'metamorphoses of science' (as compared to classical physics) – metamorphoses in search of the human dimension – including human intelligence.

In general, it would appear that the tendency of systems towards increased complexity is inevitably associated with the unpredictable, as is also the case in certain processes of 'recursive' numeration which become increasingly less predictable the longer the sequence.

For a stationary state, the condition of minimum production of entropy guarantees its stability, and for a stationary state related to linear processes or, at any rate, close to equilibrium, the condition of the deviations of the fluxes (J) and the forces (X) in relation to the respective stationary values, calculated to guarantee stability, is defined as $\Sigma_r \delta J_r \delta X_r \geq 0$. It is not always sure that such a condition will occur in non-linear states or in states far from equilibrium. A system characterized by irreversible processes of this latter type may be subject to fluctuations involving the entire system after making it unstable; thus, in the end it may also reach a more ordered state than the previous one (evolution), producing a reduction in entropy.

Biological organisms can be described by the thermodynamics of irreversible or non-equilibrium processes, and the fluctuations operating within them include mutations. All this implies, on the one hand, that one cannot talk about determinism, because an irreversible process is also an indeterministic process (Lindsay, Evans): on the other hand, it implies that the sensitivity to fluctuations may indeed promote the transition to more complex states, and thus evolution, but it also promotes transition to states in which the overall dissipation function is compromised, leading to degradation or to the end of the system. To be able to use the thermal energy of the environment is biologically important, but if the heat is excessive, burning and death may easily ensue.

The conscious activity of the human mind is interpreted as a manifestation of electrical activity due to biochemical processes. The variations in the activity of the neurones combined with secretion and receptor binding of the various neurotransmitters in the synapses demonstrate that the cerebral functions are phenomena which take place far from the condition of thermodynamic equilibrium, and in such circumstances the production of entropy may lead to an indeterministic character of cerebral function corresponding to free will.

Now, the problem is whether a material system in a state of relatively unstable equilibrium can freely 'decide' how to orient this instability. If the system is material, it has to expend a certain amount of energy (Ehrenberg) to orient the instability, which should hypothetically correspond to the exercise of the act of free will, i.e. it should *want to expend energy*, and thus the free act could be anterior to, and different from, the orientation of the instability itself, as well as from the related expenditure of energy. This involves recourse to the infinite. The operation of acquiring information necessary for the decision via physical means of perception by a physical system endowed with an observation and memory capability necessarily involves a cost in terms of production of entropy. This can be avoided only if a system is available which does not require physical means, that is to say expenditure of energy in the act of observing and deciding; what is needed is an immaterial intelligence which the physical machine does not possess and which the physical machine is not. Man is free insofar as he is not a machine, or he is not free.

The connection between mind and quantum-mechanical wave function has been intensively studied as a possible means of control of physical reality on the part of will (Walker, Wolf). When we say that we capture the possibilities by observation and by appropriately 'collapsing' the wave function, etc., the problem remains as to what this *we* is. If it is our mind which is capable of acting on the brain, it must be distinct (Eccles); neither can it be identified with the software, which is merely syntax without semantics (Searle).

Calculating reason, which uses logic as its cognitive method and the neuronal connections as its structural support, appears to function in terms of a fairly mechanistic form of relationships between causes and effects, whether we analyze the logic or the hardware. No freedom is therefore conceivable in such contexts. To have freedom, both the cognitive method and the related substantial substrate must be *distinct* from the logic and from the calculator. Logical and ontic incompleteness and indeterminacy, in terms of the mind/brain system, mean that neither the software nor the hardware can be adequately described as systems of inferential and causal relationships of the classical type. The complex of logical and ontic indeterminacy thus guarantees scope for freedom. What is certainly necessary is that the material instrument of free will should not be rigorously deterministic, with the result that it must be prone to error and degradation, or in other words, to pain.

We have said that the exercise of freedom by Man requires a measure of indeterminacy inherent in the matter of which the subject is composed, namely atomic indeterminacy, the fluctuations of systems far from equilibrium, the effects of which, moreover, are represented by the progress of evolution via the mechanisms of replicative *error* known as genetic mutation, and it is precisely this error which is allowed with causal freedom within the context of the indeterminacy of matter. Yet error means pain; progress occurs through approximation, error and pain.

The root of pain, from the biological point of view, is thus related to that of the use of freedom, since pain represents the high price which the matter of this universe has had to pay in order to be predisposed for the construction of free beings. The universe is full of beauty and full of pain. If we want the beauty of predisposed forms together with our freedom, we must accept the corruption of those forms and pain. By virtue of the anthropic principle, we may say that the universe, compatible with free will and aesthetic delight, must be a place of pain and death.

The advent of the spirit, as the evolution of matter towards spirit, is part and parcel of the movement of being in the process of becoming, implying the possibility, in principle, of the surpassing of the self which breaks the bounds of the essence, an essential, active transcending of the self, aspiring towards the absolute being upon which it depends, that is to say aspiring towards the very life of the Absolute – self-possession, freedom and truth. This is the sense of the cosmic process of which Man is a part, finding therein the sense of his own existence and his own dignity, and the meaning of his suffering.

Bibliography

Aristotle. *De Anima* II, 1, 413a.

G. Blandino. Una nuova forma di critica al darwinismo. *Per la filosofia* 4, 84–90 (1987).

G. Caglioti. *Simmetrie infrante* (Clup, Milano, 1983).

P. Churchland. *Neurophilosophy* (MIT Press, Cambridge MA, 1986, p. 2).

D.P. Cruikshank, R.M. Brown. Organic Matter on Asteroid 130 Electra. *Science* 238, 183–84 (1987).

P. Debray-Ritzen, in: C. Chabanis. *Dio esiste? No rispondono* … (Mondadori, Milano 1974, 127).

M. Delbrück. Body and Soul. *Science* 168, 1313 (1970).

M. Delbrück. *Mind from Matter* (Blackwell, Palo Alto, 1986).

J.C. Eccles. Brain and Free Will, in: *Consciousness and the Brain* (eds. G.C. Globus, G. Maxwell, I. Savodnik: Plenum Press, New York, 1976).

W. Ehrenberg. Maxwell's Demon. *Sci. Amer*. 217 (5), 103–110 (1967).

M. Eigen. Self Organization of Matter and the Evolution of Biological Macromolecules. *Naturwiss*. 58, 465–532 (1971).

D.H. Evans. Free Will and Entropy. *Nature* 317, 762 (1985).

S.W. Fox. *The Origins of Prebiological Systems and their Molecular Matrices*. (Acad. Press, New York–London, 1965).

S.W. Fox, K. Dose. *Molecular Evolution and the Origin of Life* (Freeman & Co., S. Francisco, 1972, 139).

K. Gödel. Uber formal unentscheidbare sätze der Principia Mathematica und verwandter systeme I, in: *Kurt Gödel Collected Works* (eds. S. Feferman *et al*.: Clarendon Press, Oxford, 1986, 144–94).

D.R. Hofstadter. *Gödel Escher Bach, an Eternal Golden Braid* (Basic Books, New York, 1979, 580–85).

F. Hoyle quoted by P.C.W. Davies. *The Accidental Universe* (Cambridge University Press, 1982, 117).

D. Kenyon, G. Steinman. *Biochemical Predestination* (McGraw-Hill, New York, 1969).

E. Laszlo. *L'ipotesi del campo* Ψ (Lubriana, Bergamo, 1987).

R.B. Lindsay. Physics: to what extent is it deterministic? *Amer. Sci*. 56, 93–111 (1968).

J.R. Lucas. *The Freedom of the Will* (Clarendon Press, Oxford, 1970).

P.L. Luisi. Order and Disorder in Biological Macromolecules. *Studies in Organic Chemistry* 10, 119–37 (1981).

J. Maritain. *Il contadino della Garonna* (Morcelliana, Brescia, 1973, 166).

J. Monod. *Le hasard e la necessité* (Seul, Paris, 1970, 194–95).

P. T. Mora. Urge and Molecular Biology. *Nature* 199, 212–219 (1963).

I. Prigogine, I. Stenjers. *La Nouvelle Alliance* (Gallimard, Paris, 1979).

K. Rahner. *Science Evolution et Pensée Chretiénne* (Desclée de Brouwer, Paris, 1967, 111).

E. Schrödinger. *Che cos'è la vita?* (Sansoni, Firenze, 1947, 87).

J.S. Searle. Menti, cervelli e programmi, in: *L'io della mente* (eds. D.R. Hofstadter, D.C. Dennet: Adelphi, Milano, 1985).

C.E. Shannon. A Mathematical Theory of Communication. *Bell Syst. Tecn. J*. 27, 379–623 (1948).

R. Sorabji. *Time, Creation and the Continuum* (Cornell University Press, New York, 1983 c. X).

St Augustine. *De Libero Arbitrio*.

G. Steinman. Non Enzymic Synthesis of Biologically Pertinent Peptides, in: *Prebiotic and Biochemical Evolution* (eds. A. Kimbal and J. Oro: Elsevier, New York, 1971, 31).

S. Sutherland. Science and Faith in Mind. *Nature* 290, 167 (1981).

A. Tarski. Truth and Proof. *Sci. Amer*. 220, 63–77 (1969).

R. Thom. *Stabilité Structurelle et Morphogénèse* (Inter. Editions, Paris, 1977).

H. Von. Ditfurth. *Wir sind nicht nur von dieser Welt* (Hoffman und Campe, Hamburg, 1981).

E.H. Walker. *Psychoenergetic Systems* (Gordon, New York, 1979, v. 3).

J.D. Watson, N.H. Hopkins, J.W. Roberts, J. Steitz, A.M. Weiner. *Molecular Biology of the Gene* (The Benjamin/Cummings Publ. Co., Menlo Park, 1987, 1104).

G. Webster. Natura e scopo dell'analisi strutturalista in biologia. *Riv. Biol. – B. Forum* 80 (2), 275–90 (1987).

J.A. Wheeler quoted by G. Gale. The Anthropic Principle. *Scient. Am.* 245 (6), 154–71 (1981).

J.J. Wicken. *Evolution, Thermodynamics and Information* (Oxford University Press, 1987).

F.A. Wolf. *Mind and the New Physics* (Heineman, London, 1985).

M. Zatti. The Bio-Bang. *Riv. Biol.–B. Forum* 80, 79–100 (1987).

Metaphysical Outlooks in Physics and the Anthropic Principle

N. DALLAPORTA

Dipartimento di Astronomia dell'Università – Padova

S.I.S.S.A. – Trieste

1. Introduction

Up to a few decades ago, physics and cosmology were considered on the whole as an ensemble of observational, experimental and theoretical laws, which have to be taken as they are, and can be reduced to a minimum number of general principles, without asking their reason for being such.

For my part, I think that an enormous advance was made in the domain of conceptual science, when the question was raised concerning why the world is as it is; this step was the starting point for a search of deeper seated reasons ruling scientific research in respect to what had been looked for before.

In short, leaving presently aside the somewhat related question why space has just three dimensions and no more or no less, which we do not propose here to consider notwithstanding its interest, the type of analysis based on the **why** instead of on the **how** has focussed, as principal factors on which depend the main characteristic properties of the world, two essential groups of data or phenomena:

(1) the global aspects and the initial conditions of the cosmos ruling the course of its history;
(2) the properties of the fundamental fields and the related values of their interaction constants.

What is now called the anthropic principle is nothing else but a statement mainly concerning the connections between the preceding two types of general parameters and the possibility for life to grow in the whole universe.

There are two main levels for expressing such connections: the weak form of the principle, stating that a number of observational data could not have been found much different from what they are, because otherwise nobody would be present in the universe to witness them; and the strong form, stressing a more compelling link: the main parameters of the physical world are such in order to allow life, and hence man, to be formed.

The number of connections showing that even very small changes in the values of several of these parameters should prevent life from arising is overwhelming; and the excellent book of Barrow and Tipler probably represents the most extended summary concerning everything that has been stated on this subject. The aim of this talk is not to present a review of these statements – a task which would require almost as long a time as for reading the book itself – but rather, assuming them as granted, to discuss the several meanings put forward for them, which obviously present the philosophical tendencies of their proposers, and to compare their relative likelihoods.

The simplest way of tackling such an aim, in order to reduce the whole question to its main elements, consists perhaps in referring to those theoretical approaches, directed towards maximum synthesis, intended to reduce as much as possible the total number of independent parameters on which the properties of the cosmos appear to depend. I know of two such approaches, each of which tries to collect under a single leading idea one of the two groups of parameters previously quoted; they are:

(I)   Inflationary cosmology, looking for a unification of the initial conditions in the universe; and

(II) Grand Unification Theory (GUT), with equal aim in relation to the different types of particle interactions.

As a first goal, I now intend to stress that these ideas leading towards synthesis are however fundamentally different and in a sense opposite.

2. Inflationary Cosmology

In its standard formulation by Guth, Linde, Albrecht and Steinhardt, inflation is a tentative unification of the three main problems of initial cosmos, i.e.:

(a) **the horizon problem** – or why in portions of space which up to recent times have been causally disconnected, properties such as the blackbody background radiation temperature are still the same in all directions, as if in the past there had been a causal connection between them;

(b) **the flatness problem** – or why the model of the universe appears to be the flat-space one, with nearly critical density;

(c) **the fluctuation problem** – concerning the amplitude of the density irregularities allowing the later formation of cosmic protostructures.

These questions are reconducted to the existence of a phase transition of matter at about 10^{-35} s after the Big-Bang. Accordingly, the very peculiar properties of the universe are shown to arise at such an inflation time (10^{-35} s), while the previous situation could have been of any kind, which means that a wider set of initial conditions in any case would have led to the precise configuration of the world we do actually observe. Such a conception is basically similar to

Misner's chaotic universe: in both cases, after a chaotic beginning, order and regularity appear only later, as a consequence of a uniformizing factor embedded in the nature of things.

The chaotic character at the start is still more strongly stressed in later outlooks, having replaced the phase transition model. According to Linde's proposal, inflation itself has to be considered as chaotic, in the sense that its features will depend on random fluctuations for a primordial field, different from place to place, and the universe will finally exist in a variety of bubbles, each of which will have expanded according to different rates, so that all possible connections and correlations between the different parameters are likely to be found. What we call our universe is supposed to be just one of these bubbles, having just by chance been endowed not only with the right initial cosmological conditions, but also with the right interaction constants among fields, allowing life to develop.

A third step in this main advance of the chaotic outlook is represented by the idea of an infinite number of universes considered as a complete set, including all the infinitely numerous possibilities of all different combinations of the constants, a larger set in respect of the number of effectively realized bubbles of Linde's chaotic inflation model; and this could bring some differences in the deductions from both.

3. Grand Unification Theory

Let us now turn to the second general approach, related not to cosmology, but to particle physics, aiming to reduce the different fields and related interaction constants to a single entity from which all of them might be reasonably deduced. The idea is just a further implementation of the achievement of Maxwell in unifying electric and magnetic forces, and more recently of Weinberg, Salam and Glashow combining into a single field electromagnetic and weak interactions. As all interactions are ruled by symmetries, one has to look for more and more extended and comprehensive ones, wide enough for including all subsystems now observed, appearing as different subgroups when the larger symmetry, valid at very high temperatures ($T \sim 10^{15}$ GeV), is broken at lower ones. Of course, the all-unifying symmetry should be described by fewer, or even a single parameter. The different parameters, measured in present conditions, should in principle be deducible from them.

This is enough for understanding that in this case the unifying principle is just the opposite in respect to inflation cosmology. The simplest way of looking at it is to take it as an *a priori* idea, expressed by the largest symmetry, ruling the basic matter and imposing on it its peculiar forms, appearing as the different fields and structures of the world.

I would like to mention that I do not feel competent enough to deal with the approach pursued, in opposition to most cosmologists, by Hawking and collaborators; their basic idea consists of looking for solutions of anisotropic type for the universe, which seem to rely on very special conditions. This line in some sense is ideally more related to the GUT assumption of particle physics than to the usual chaotic approaches in cosmology.

So, while according to the leading point of view in cosmology, the universe is ruled by pure chance, for particle physics it can be seen as an *a priori* project; and a project cannot be thought about without a planner whose mind has conceived it: so, the ultimate decision concerning the real nature of things has to be taken between these two alternative views.

4. Recourse to Metaphysics

Now, the idea of a planner of the universe is of course not a physical, but a metaphysical idea. And I consider that the main impulse leading towards the opposite chance or chaotic assumption has been mostly pursued with the purpose of avoiding any invasion of metaphysics into the physical domain. The aim of my talk is just to show that such a purpose has not been achieved, and that in reality the chaotic assumption has to rely on metaphysics at least as much as the planner assumption, and in a much more complicated way than the direct recourse to God.

In such a tentative analysis I will take the word *metaphysics* in its etymological sense as something lying beyond the range of physical research, with only indirect references to the deeper aspects of reality implied by it. I will therefore begin by discussing the different types of doubts inherent to all early universe models following from different kinds of extrapolations.

The first and most straightforward one relates to how much of physical reality we might expect to find in all kinds of enquiries down to the Planck dimensions (10^{-33} cm, 10^{-43} s, 10^{93} g/cm^3 and so on). At present we cannot expect to succeed in measuring directly phenomena related to such dimensions in any thinkable future. Therefore we might first ask what amount of reality in the ordinary physical sense can be reasonably attributed to the very initial phases of the Big-Bang picture: I would like to propose the following answer.

Let us assume that from the conditions postulated as valid in the early situations we are able to deduce the theoretically expected values in our time of two quantities A and B; and that in fact we observe their correlation as predicted. To my mind, this would be a proof that the scheme is coherent and logically consistent, and could therefore represent reality; but I would not conclude that it is in fact reality in the ordinary physical sense, when everything implied in it could be directly measured.

In fact normal caution and experience should warn us that in almost all important advances of physics, straightforward extrapolations of laws valid in a given domain into a dimensionally different one have generally turned out to be incorrect; and I would add that several fundamental constants of physics act as landmarks indicating the border of validity of the laws considered: the limits of Newtonian mechanics in respect to velocity are marked by the velocity of light c, and those of classical physics in respect to small masses by the Planck constant h. This constitutes a strong indication that generally we cannot extend to a very different dimensional scale some given physical laws without modifying on the same occasion these laws themselves. Of course it can be said that such a warning has been taken into account by the concept of the Planck limit itself, introduced as the border beyond which interference occurs between general relativity and quantum mechanics; this fact would require the quantization of the gravitational field, an enterprise not yet achieved. However, I would like to stress that this is an expected modification due to the superposition of three already well-known borders, c, G and h; instead in the previous considerations, the borders separately due to c and h leading to the new domains had turned out as quite unexpected in respect to earlier physics. Therefore I feel that in the extrapolations towards the Big-Bang, other unexpected borders are likely to appear; and, to my mind, this adds some touch of metaphysics to the currently adopted Big-Bang picture.

A second kind of extrapolation on which I bear some doubts concerns the absoluteness granted to the principles of quantum mechanics, assumed to hold in conditions totally different from those valid in the physical surrounding for which quantum mechanics has been constructed. As far as I know, the indetermination relations have been deduced from conceptual experiments requiring an observer for their accomplishment. What then is meant exactly by a wave function related to the whole universe undergoing quantum fluctuations, when there is no observer to cause or to witness indetermination unless we refer directly to God for playing such a role? I am afraid – although I might be wrong – that such exotic uses of quantum mechanics, so different from the standard ones, might hide some kinds of extrapolations of the normal concepts into rather unphysical – and therefore metaphysical – regions.

If the different bubbles postulated by Linde's inflationary cosmology are a consequence of this initial determination, the previous doubts might cast corresponding ones even on their possibility. And if, in spite of such doubts, we still wish to rely on this bubble picture, we are faced with the two following alternatives:

(a) Should the universe be finite, and therefore the total number of bubbles also finite, and their characteristic parameters formed at random, then the

probability of any of these bubbles having exactly all the required parameters for producing our world, and life in it, would be extremely low.

(b) Should we instead consider, in an infinite universe, the number of bubbles
     as forming a complete infinite set of mini universes, then certainly one of
     them should possess the required characteristics. As it is however nor-
     mally considered that the expansion velocity of the bubbles is superlumi-
     nal, then no physical connection, as far as the horizon grows, could even
     be established between bubbles; so these mini universes would constitute
     a purely metaphysical assumption, with no hint of testable, and therefore
     physical reality in them.

My general conclusion is that in order to justify the various previous
assumptions current in present day cosmology, it is necessary for each of them
to build a frame of metaphysical postulates much more involved and artificial
than the opposite straightforwardly metaphysical view of a universe built
according to an *a priori* plan, requiring a planning Intelligence adequate to
have conceived it.

Thus it appears that a recourse to metaphysics is inescapable, whichever
view we choose for explaining the physical world. This being so, I am naturally inclined, as a physicist, used to considering the simplest theory as the
best one, to behave in the same way in the metaphysical domain. This is why,
according to the present analysis, I have no doubts, given the chaotic and the
planner assumptions, to give my preference to the second one, and I am also
inclined to think that the correlations pointed out by the anthropic principle are
a strong indication for a general design connecting the physical and the human
worlds.

5. Conclusions

As a final remark, I would like to analyze briefly why a large number of scientists are so much against the design interpretation of the anthropic principle, up
to the point of resorting to the unlikely postulates of the chance interpretation.
The answer is obvious: their strong opposition is due to the fact that the design
idea relies on finalism; and, since the last century, finalism has been radically
contrasted by the current scientific mentality. Now, man's normal way of
thinking and behaving is an intimate mixture of causal and finalistic reasons,
so that both these viewpoints are quite pertinent to human intelligence. One
then would like to understand why, in the building of science, its whole construction has relied only on causality, with complete rejection of finalism. I see
two main reasons for such an attitude.

The first one is that, for a long time, especially in the domain of biological
sciences, the idea of finalism has been frequently used in a rather childish and

unintelligent way, relying sometimes on dull and formalistic interpretations of biblical texts. The reaction of science to such a misuse has been strong enough to banish finalism from its domain in all its possible aspects.

However, there is, to my mind, a deeper seated reason for this attitude: science has started its own development by solving the simplest physical problems with few independent degrees of freedom, as is, for example, the two-body problem. In this case the situation is entirely solved according to pure causality: given the initial conditions, the earth around the sun and the electron around the proton will follow the same path for all eternity. Such a clearcut solution has been assumed as the ideal prototype for any physical problem and as the backbone underlying the intimate structure of the world. However, we know that any perturbation brought by a third body complicates the situation so much that no general solution can be obtained in this case; and then complication grows exponentially with the number of particles involved, so that the consideration of any multiparticle problem as the result of the sum of two-body interactions is just an idealistic view with no possibility of a straightforward verification; and a real collapse of standard determinism may be finally recognized in the occurrence in several situations of the so-called *catastrophes;* so that, on the whole, the role of deterministic causality, so clearly marked in the two-body case, in some sense becomes a Utopian point of view. Would it then be such a heretical idea to surmise that in very complex problems a finalistic point of view could work as a useful tool? Finalistic aims are evident in biological situations: would it then not be much easier, instead of trying to reconduct them to an enormous complexity of causal interactions, to consider them as a complementary outlook allowing a kind of shortcut to operate through this causal complexity towards a simple resultant behaviour of a given biological phenomenon? In a sense, finalism might appear as being the reverse of causality, obtainable from it by inverting the direction of the time, and exchanging the roles of past and future; so that, in complicated cases, the reason for some happenings could be looked for in the future instead of in the past. Complementarity might then operate in the sense that when causality is dominant, as in the two-body problem, finalism disappears; but in the opposite case of complicated situations, finalism could be dominant and causality more or less unobservable.

To my mind, the anthropic principle might represent the first significant hint towards such a kind of understanding of the universe. Analyzed in its elementary interactions, we have to refer as its leading aspect mainly to causality; but when viewed in its total complexity, the aim or the design might become overwhelming. Should this be true, the recognition of the anthropic principle should be considered as a turning point in the development of science, opening new roads towards the unknown aspects of the universe.

I wish to express my friendly thanks to Silvio Bonometto, Francesco Lucchin and Giovanni Prosperi for their several suggestions and contributions to the discussion of the different problems involved in the present subject.

Galaxy Creation – Implication for the Development of Life

HALTON ARP

Max-Planck-Institut für Astrophysik
München, F. R. G.

We all look forward to conferences like this because we hope that we will learn something new and important. A conference on the anthropic principle seems particularly attractive because the postulated intimate relationship between the nature of humanity and the nature of the universe promises us revelations about one by studying the other. But when I think in specifics I become discouraged. The understanding of mankind proceeds very slowly – I suppose because of his psychological deviousness. On the other hand we do not know very much about the universe either. In fact I am especially pessimistic about knowledge of the universe because I believe we know even much less about the universe than we think we do.

In the following discussion I am going to argue that simple consideration of images of galaxies in different optical and radio wavelengths demonstrates that galaxies are continually created. I will argue that, unlike the conventionally accepted Big Bang theory, galaxies are not created at just one instant near the beginning of the universe, but are being created now and presumably indefinitely far into the future. If this conclusion contradicts current accepted wisdom then each person must make his own choice from the evidence as to which picture is most likely to be true. But even if one continues to hold the Big Bang theory of creation one must be impressed with the fact that the arguments presented below lead to a radically different viewpoint. At the very least this represents an enormous uncertainty in our present knowledge of the universe. On the other hand if, in fact, we do have continuing galaxy formation in any way like the processes described below it would have profound implications for the formation and evolution of life in the universe. I would like to comment on some of these implications at the end.

1. The Age and Origin of Galaxies

What can we tell about the age of a galaxy by looking at it? Really all we need are photographs in two separate colours, say blue and red. We can then look at the proportion of blue stars, which are hot, fast burning and short lived, compared to the proportion of red stars, which are cool and generally old. We see a gradation in kinds of galaxies, from those composed almost entirely of old stars to those composed almost entirely of young stars. The most natural conclusion would be that the latter galaxies are younger. But that would contradict a central assumption of the Big Bang: that all galaxies formed 16 billion years ago shortly after the initial explosion. In order to have galaxies filled with young stars today the conventional theory must resort to an added, complicating assumption. It must suppose that this particular kind of galaxy formed long ago but that it formed stars within it very slowly. In fact the galaxies now clearly filled with young stars must just now be undergoing a sudden burst of star formation at the moment at which they are being observed.

But there is direct observational evidence that this cannot be so. The reason is that stars are currently believed to condense out of large clouds of gas. If stars were currently condensing out of such clouds the clouds should be visible to, for example, radio telescopes which are very sensitive to hydrogen and molecular gas. There should be galaxy sized clouds floating around in space which have not yet started, or are just about to start, forming stars. Such clouds are obviously not detected.

For example the compact, blue galaxy called 3C 120 shows a young stellar spectrum (Arp 1968; Baldwin *et al*. 1980). (The optical appearance of 3C 120 can be seen within the central radio contours mapped in Fig. 1.) There are no star forming clouds of gas and dust visible in its vicinity. From where do the abundance of young stars in this system come? An obvious sign which points to their origin are the short, luminous jets which emerge in opposite directions from its compact, active nucleus. That nucleus is so brilliant – and a source of radio waves and X-rays – that the galaxy was originally called a quasar (see Arp 1968; 1987). The jets mean that material is being ejected from this nucleus. Where does the young material in the outer parts of the galaxy come from? Obviously the nucleus.

If we look in Fig. 1 at the radio map of the region surrounding 3C 120 we see long radio extensions far out beyond the optical jets. These are typical emissions from ionized material moving along a magnetic field; radio wave emitting material which is commonly accepted to have been ejected from the active nucleus. The important point to realize, however, is that such behaviour is typical of hundreds to thousands of radio galaxies. It is simply a characteristic of galaxies that they intermittently eject material far out into space. If we actually heed the observational evidence we see immediately that galaxies are typically not just quietly rotating under gravity in a near equilibrium condition.

This particular galaxy, 3C 120, is exemplary in another way. Looking out along the northwest jet one sees a small galaxy of peculiar, non equilibrium

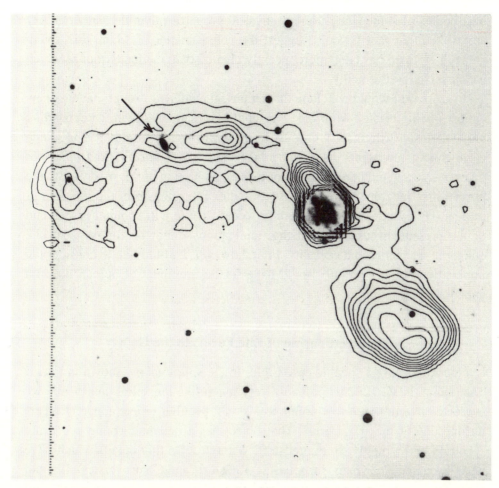

Fig. 1 The quasar-like galaxy, 3C 120, is shown inside the radio emission contours. The optical jet from the active nucleus points out along the direction of ejection of radio material. The arrow points to a young stellar population, non equilibrium form galaxy with a +5000 km s^{-1} redshift relative to 3C 120.

form (marked by an arrow in Fig. 1). Its spectrum shows a strong concentration of early type (young) stars. So both its form and content tell us that it could not have existed for very long. But from where has it come? Tracing back the path of the ejected radio material leads us directly to the nucleus of 3C 120.

Now at this point it could be argued that the ionized gas which we know is ejected from galaxy nuclei just recombines, condenses and forms this new galaxy in more or less the usual fashion in which it is believed stars form (see Arp and Sulentic 1988). But it is also possible that lumps of material could be ejected which then differentiate into stars in a somewhat different process. Since the young companion galaxy has what seems best interpreted as an intrinsic redshift of about 5,000 km s^{-1} I would favour the hypothesis that compact bodies of somewhat different physical properties than the parent galaxy are ejected. (In fact I would suggest the actual matter of which this young galaxy is formed has been more recently created than the matter out of which we ourselves are formed.)

In any case I would conclude that empirical evidence strongly favours new galaxies being ejected from old. This could be reconciled with an expanding universe which continually unfolds from within itself – a universe in which each galaxy represents its own 'mini-bang' (see *Quasars, Redshifts and Controversies*, cited in references for further discussion of this model). But the possibilities inherent in newly created matter emerging from the interiors of present day galaxies are even more provocative. We will comment at the end of the present discussion what the implications might be for subjects like quantum mechanics and causality. First, however, I would like to consider the implications such a picture would have for development of human life in a galaxy.

2. The Relationship of Life Forms to the Galaxy

In order to discuss a specific case I refer to the spiral galaxy shown in Fig. 2. That spiral, NGC 1232, is probably very much like the spiral galaxy that we ourselves live in. An arrow points to a small, peculiar object apparently perturbing one of the spiral arms. That object has a peculiar, young appearing spectrum and a highly excess redshift. Without repeating details which have been discussed elsewhere (see Quasars, Redshifts and Controversies), I would argue that this is a newborn galaxy, and that it has just recently emerged from the nucleus of its parent galaxy along the spiral arm which represents an ejection trajectory back to that nucleus.

For the moment let us concentrate on the spiral arms as ejection trajectories from the nucleus (additional references can be found in 'Persistent Problems of Spiral Galaxies' cited in references). We know that spiral arms such as those in our own galaxy represent a concentration of gas, dust, molecular clouds, cometary and planetary bodies – in short all the chemical constituents that make up our own life-populated earth and out of which our earth was presumably formed. But how formed? As is customarily believed, has this material been just hanging around from the beginning of the universe and forming stars and planets from time to time? Or is it material which was recently transported from the centre of the galaxy and in which the seeds of new worlds are sprouting?

Equally intriguing, perhaps, is the question: 'What exactly is this material which is concentrated in the vicinity of spiral arms – the only place where we know for certain that partially intelligent life forms exist?' Not so long ago it

Fig. 2 The spiral galaxy NGC 1232 pictured here is probably quite similar to our own Milky Way Galaxy. The arrow points to a peculiar, young galaxy of high redshift which appears to perturb the clouds of star forming gas and dust at that point in the spiral arm.

was thought that the interstellar clouds were simply gases, with perhaps some simple carbon and silicon based dust grains. But now it is becoming rapidly evident that there are huge clouds of extremely complex molecules. These are the molecules out of which the living matter we know on earth is made. They are carbon, hydrogen, nitrogen and oxygen compounds. Fred Hoyle and Chandra Wickramasinghe have reported that spectra of bacteria pulled out of an English river match almost exactly the infrared spectra of interstellar clouds. For over 70 years diffuse interstellar bands have been observed in the optical spectra of interstellar clouds. It is clear these clouds break down into CH and CN in the vicinity of hot stars (Krelowski 1988). Can it be that the failure for more than two generations to identify the spectral features from these interstellar clouds is because they are in a molecular form as complex as Hoyle's virus/bacterial organisms?

Interplanetary molecules were recently sampled in the satellite fly-by of comet Halley. Some researchers argued that they saw insufficient phosphorus in the molecules to support life as we know it. But do we know how much phosphorus is needed for all forms of life in all stages? Is it not more striking that all the myriad similarities to life molecules exist than that some details are different? From the long perspectives of several decades there is considerable, and perhaps growing evidence that the universe we live in is more of an organic universe than an inorganic one. Is this not a congenial result from the viewpoint of the anthropic principle?

Before we go back to the question of from where this material originated we should make one or two more remarks about life forms and galaxies. If we define life forms operationally as objects which move from internal forces and reproduce, then we are faced with the question of scale. Our scale of experience is limited to not much larger than the human being. Would we recognize complex, non mechanical behaviour on an interstellar scale? Would a bacterium recognize man as an intelligent life form?

On an even larger scale, if galaxies can move from internal forces and produce smaller galaxies, do they fit the definition of an inorganic entity? If the universe does obey an anthropic principle, is it selected by the needs of viruses, humans or galaxies?

3. Creation of Matter and Communication in the Universe

If galaxies create new galaxies in the ongoing process postulated in section 1 of this paper it is natural to ask the following question: 'From what underground spring does this nucleus flow?' In the conventional picture old matter in a galaxy would fall into an active nucleus and be blown out into space in the ejections we observe. But in the strongest cases of galaxies feeding on themselves we would soon see emaciated, skeletal galaxies lying around. We observe nothing of the kind, the largest galaxies we know tend to be the most active, full of stars both old and young and showing no signs of diminishing mass.

In the nucleus of many galaxies is observed, however, a very small, very active centre. In these the mass/energy density is supposed to be very high. From there fountains of material are observed to arise. Could matter be 'created' at this point? To consider such a possibility has not been fashionable in physics and cosmology. But a few years ago Alan Guth, in attempting to join current knowledge of particle physics to a Big Bang model of the universe, helped to develop the so-called inflationary theory. He noted that there could be fluctuations in what physicists call the material vacuum where whole new universes could be born out of effectively nothing. Amazingly, after 60 years the idea had finally occurred to someone that if the universe had been formed in a Big Bang, then there was no reason that it could not happen again!

Most recently the Russian astrophysicist Andrei Linde has developed a theory of many universes being created throughout space but not within light travel contact with each other. Ilya Prigogine and collaborators have pointed out that a term could be introduced in the Einstein matter/energy tensor that allowed for matter creation. Stephen Hawking in his recent best selling book *A Brief History of Time* proposes creation of matter at the edges of a black hole. But are these really new ideas? More than 20 years ago Bondi, Gold and Hoyle proposed a steady state universe in which new matter was created in order to keep the density steady in an expanding universe. Fred Hoyle called his term in the general relativistic equation the 'C Field' where 'C' stood for creation. But even much earlier the eminent physicist Paul Dirac had proposed two kinds of matter creation: one in empty space, and the second in the presence of matter already present in the universe. Where would the greatest previous concentrations of matter occur? Obviously in just the galaxy nuclei which we have observed to be the densest and also the most active in erupting material out into space. The Russian astrophysicist Zeldovich remarked shortly before his death that theories of matter creation were like 'sleeping beauties' now ready to be reawakened. But perhaps we would all make progress faster in this world if we allowed our 'theories' to be guided more by the observations.

What do the observations tell us about creation of matter? First of all, if it takes place in the tiny central points of galactic nuclei, for example in nuclei which are resolved by very long base line interferometry (a few hundredths of an arc second), the material must originate in a very small volume. (If these objects are much closer than their redshift distances, where physical association places them, and where their expansion velocities are no longer superluminal, then the central volumes are even much smaller.) It is unreasonable to suppose great mass is introduced at these tiny points. Moreover strong mass would produce very large dynamical perturbations which are generally not seen. So we would *induce* that the matter was created in a low mass state. This is, however, exactly the logical behaviour one would predict for newly created mass. A particle can only know how much mass it has by exchanging gravitons with the rest of the universe. A newly created particle (created from a non local field) can only communicate with a very small portion of the universe. Therefore we would expect newly created material to start off from nearly zero mass and grow to more normal mass as it ages. It turns out that there is a perfectly rigorous and complete physical theory which permits the mass of particles to be a function of time. It is the Hoyle–Narlikar theory of conformal gravity. Physicists have said we do not need this theory because our present theory tells us everything. But the observations say to me that the old assumption that mass is not a function of age is observationally untenable.

A highly controversial, but to me decisive, observation is that intrinsic redshifts of extragalactic objects vary with the epoch of their creation. (Lower mass particles in an atom would perforce give redshifted radiation.) Many proofs are now available that very high redshift objects are relatively close by in extragalactic space, associated with much lower redshift objects (see *Quasars, Redshifts and Controversies*). The most recent proof is that quasar redshifts are periodic, i.e. certain preferred values of quasar redshifts are observed such as $z = 0.60, 0.96, 1.41, 1.96$ (Arp, Bi, Chu and Zhu 1989). If quasars were at their redshift distances this would violate the 'no preferred position' (cosmological principle) because we would be at the centre of concentric shells of quasars. Therefore the quasars must be relatively nearby, not out at the edge of the universe where their conventional redshift distances would place them.

The redshift periodicity tells us something additional, however. If material starts from a zero mass state it starts from the dimensional domain of quantum mechanics. There we would *expect* quantization of properties such as mass and redshift. If the quantization of properties is not implanted at this stage there is no way quantization could be later induced in a macroscopic body. So I would regard the observational result of quasar periodicity as a proof of the recent creation of high intrinsic redshift, nearby quasars. I would propose that these quasars evolve into the young galaxies which we see continually arising in the universe.

Finally we can ask how such continuous creation in the universe would affect out views of the origin and future of life forms in the universe. I would argue that the universe must be defined as everything that is now or in the future detectable. In that sense 'other universes' cannot exist which inject new matter into our universe. If matter appears where none was before it must have come from some other place in the universe. But I do not speak here of teleportation which violates the known rules of classical physics. I instead suppose that previous to its materialization at a certain point in spacetime it existed as a diffuse field or potential. That is, it pervaded the entire universe.

This is the observational conclusion for quantum mechanical particles on a microscopic scale. Again we are led back to zero mass. Quantum mechanical materialization of new matter, new galaxies, representing a transition from a non local to a local state in the universe.

Now the presence and evolution of life forms, whether they have hands or tentacles, whether they are individuals or swarms of communicating entities, whether they are of small or large scale, may be important or trivial to this process of galaxy creation in the universe. But if the matter out of which life is formed, has been spewed out of a nucleus from some diffused state pervading the universe, then there is some physical relationship of life to the universe as a whole. The operational definition of human life as we know it seems to be to organize information and effect change. Possibly we have the ability to carry on this process for an indefinitely long time. In that case we, or some other life forms, may have an important role in the universe. In that case the anthropic principle could represent a law of causal physics like gravity or the flow of time.

General References

Quasars, Redshifts and Controversies by Halton Arp, publ. Interstellar Media, 2153 Russell St. Berkeley 94705, Calif.

'Persistent Problem of Spiral Galaxies' by Halton Arp, *IEEE Transactions on Plasma Science*, Vol. PS-14, No. 6, Dec. 1986.

'The Red Shift Controversy – Further Evidence Against the Conventional Viewpoint' by Halton Arp, *Comments on Astrophysics* 1988, Gordon and Breach, in press.

Specific References

Arp, H. 1968, *Astrophys. J.* **152**, 1101.

Arp, H. 1987, *J. Astrophys. Astr. (India)* **8**, 231.

Arp, H. and Sulentic, J.W. 1988, 'The Properties of NGC 2777 – Are Companion Galaxies Young?', *Max-Planck-Institut für Astrophysik* Preprint No. 388.

Arp, H., Bi, H.G., Chu, Y. and Zhu, X. 1989, 'Periodicity of Quasar Redshifts', *Max-Planck-Institut für Astrophysik* Preprint No. 427.

Baldwin, J.A., Carswell, R.F., Wampler, E.J., Smith, H.E., Burbridge, E.M. and Bokensberg, A. 1980, *Astrophys. J.* **323**, 4.

Krelowski, J. 1988, *Pub. Astr. Soc. Pacific* **100**, 896.

Some Theological Reflections on the Anthropic Principle

GEORGE V. COYNE, S.J.
Vatican Observatory
Città del Vaticano

Abstract The Anthropic Principle has made a significant contribution to bringing back into harmony the discordance that has existed since the time of
Galileo between the scientific and the humanistic cultures. In the light of this,
and while respecting the methodological independence of science, philosophy,
and theology, I wish to explore how the Anthropic Principle as a scientific
conclusion may be an incentive to theological reflection.

1. The Divorce Between Science and Humanism

One of the principal issues at stake in the Galilean controversy was the
methodological independence of science and theology. Galileo addressed the
issue in a very specific way in his *Letter to the Duchess Christina*.[1] In a sense
both sides of the controversy were at fault, Galileo for not appreciating the
hypothetical character of the Copernican model of the universe and theologians for failing to realize that Scripture was not teaching science. While theologians did not respect the complete autonomy of science as a way of knowing, the scientists, characterized by Galileo, respected neither the tension
between hypothesis and truth in scientific methodology nor the difficult passage from one to the other through both an inductive process from observations and a deductive process from mathematical physics.

The result of the controversy, although not deliberately nurtured by either
side, was a divorce between nature and the human person. In fact, science was
intent upon removing the element *human person* from its methods of investigation so as to preserve the objectivity characteristic of the sciences. The two
cultures, science and humanism, went their separate ways with little discourse
between them.[2]

2. The Anthropic Principle

The Anthropic Principle, first enunciated as such by Carter,[3] has served to
mend that separation. Other papers in this publication will have enunciated the

conclusions of Carter in more detail. For my purposes let me simply summarize Carter's conclusions. The emergence of human civilization has required an extremely fine-tuned combination of physical constants and laws of nature from the very beginning of the universe in a primordial very dense, very hot state and throughout the evolution of the universe. The so-called weak version of the Anthropic Principle simply sees this as an observational effect and, in fact, it would be more meaningful to call this version the Observer Principle. We observe the universe to be fine-tuned because if it were not fine-tuned we would not be here to observe it. In this version the Anthropic Principle is simply one of the many selection effects that observers must cope with in evaluating the data they obtain from observations. On the other hand, if one proceeds beyond the recognition of the fine-tuning as a selection effect and dares to ask the question *why*, one enters into the realm of the so-called strong Anthropic Principle, whereby one seeks to explain the origins of the fine-tuning and the reasons for the precise values of the many fundamental constants and for the laws of nature. As we shall see, it is difficult to do this without entering into a dialogue with philosophical and theological considerations. Up to the present the fundamental constants have, for the most part, only empirically determined values. There is no fundamental physical-mathematical model from which they can be derived. There is no unified theory which explains all of them. They are simply found from observations to have the values that they have and even a slight change in them would exclude the evolution of the universe to human civilization.[4] By using the words Anthropic Principle to denominate his conclusions Carter obviously insinuated some kind of finality in the evolution of the universe leading to human civilization. Whatever might be the cosmological model used to explain that finality, be it real or apparent, we are inevitably invited to philosophical and theological reflections.

3. Response to the Anthropic Principle

Since the Anthropic Principle, at least in the strong version, leads to investigations which strictly transcend the methodology of science, many scientists simply reject it as not susceptible of scientific enquiry. Others see it as indicating a certain intrinsic finality in nature but without reference to the origins of such a finality. Still others from a religious persuasion see it as indicative of the existence of a Supreme Being who created, among many other possibilities, a universe in which human civilization would emerge. There are finally those who, prescinding from any philosophical or theological considerations, simply reject the Anthropic Principle as of no value to science, since it can neither predict testable conclusions nor assist in the planning of research programmes leading to a further understanding of the universe. While I must concede that a predictive character appears to be lacking, the Anthropic Principle has certainly provided an incentive for research in cosmology.[5] Furthermore, and perhaps more importantly, it has reinserted the human person into the continuing search for a total and comprehensive understanding of the universe. While certain aspects of this reinsertion may transcend the strict boundaries of what are the proper object and methods of scientific enquiry, it is not obvious that all aspects may be so excluded. At any rate the Anthropic Principle certainly provides an invitation for a serious dialogue among scientists, philosophers, and theologians.

4. The Encounter of Theology and Science

As an example of how certain scientific conclusions may influence theological reflections, I would like now to investigate how the various cosmological models proposed in response to the Anthropic Principle might contribute to elaborating the religious concept of God. As an object of religious enquiry and of faith God is the supreme mystery. Nevertheless, the religious person believes that God has spoken of himself to the human race through the prophets, the patriarchs, and for the Christian believer through his Son. This self-revelation of God is found in religious traditions and in the holy books. Theology is properly speaking a science (in the wider sense of that word), a way of knowing, with its own rigorous methodology. Through such disciplines as literary analysis, philosophy, linguistics, etc. it studies religious traditions and the holy books in order to discover religious faith and the object of that faith, God. I would like now to confront the concept of God derived by theology with the Anthropic Principle derived from scientific enquiry. To be more specific, I wish to address the question: Among the various cosmological models proposed in response to the Anthropic Principle, is there one which is more consistent with the concept of God derived from theological enquiry? From the very beginning of these reflections it is necessary to establish two points:

(1) of the many cosmological models proposed there is none yet which even approaches being definitive;
(2) between cosmology and theology we are looking for consistency and not for definitive or determinative concepts or, much less, for proofs.

5. The Knowledge of God in Religion

How does the theologian arrive at a concept of God? We find ourselves immediately in an epistemological dilemma. By definition God is mystery and unknowable in himself. The only way that we can approach a concept of God is by the negative way,[6] that is by taking that which we find from our living experience to be good, beautiful, and true in ourselves and in the world about us, stripping off (denying) the imperfections that we experience and then attributing the purified attributes by analogy[7] to God. Furthermore, as I have mentioned above, God has spoken to us of himself, and so we can interpret what he has said to us through religious traditions and the holy writings, always however using the negative way. The religious traditions have been experienced by and handed down by human beings; they are, therefore, fallible. The holy books have been written, handed down, and read by human beings; they are, therefore, fallible. In order to arrive at the source, at the God who spoke, we must study in a rigorous and scientific way those traditions and those writings. We must, in other words, understand the human transmission of what God said in order to arrive at the God who is speaking. We must purify the transmission of its human imperfections.

As an example of the negative way let us now consider a fundamental attribute of God. He is free and is, in fact, the source and foundation of all freedom. From the fullness of his freedom he has created the universe and has formed the human being to his own image and likeness;[8] he has loved that creation and he chose a people to whom he sent his own Son. These are some of the fundamental ingredients of Christian belief in God. On the other hand in the exercise of his freedom God is not arbitrary; arbitrariness is a defect. To choose by whim and fancy without sufficient reason and motivation is unbecoming of God. In the Scriptures and in religious traditions God is sometimes seen in this way. One must, therefore, while preserving the primordial presentation, many times in story form, of God's freedom, purify the presentation of its negative and imperfect characteristics. In brief, one must apply the negative way.

6. God the Creator and Cosmological Models

In religious thinking the concept of God the Creator has always been open to the risk of presenting God as choosing in an arbitrary manner. The stories of the creation in the Book of Genesis are a primary example of this. However, the Genesis stories really intend to present a much more fundamental characteristic of God, namely that he is the saving and redeeming God. Genesis is really saying that the same God who saved mankind is the God who created the world, and that, in fact, his creating is a salvific act. Genesis is not at all interested in how God created the world, even though it presents stories, some of them with common origins in other contemporary cultures, to show that the creating God is loving and salvific.[9] Nevertheless, it is difficult to escape the fact that from the presentation in Genesis and throughout religious traditions there is a certain arbitrariness that creeps in to the concept of God-Creator. Let us, therefore, proceed with the task of attempting by the negative way to purify the concept of God-Creator, from a certain inescapable character of arbitrariness by confronting it with the various cosmological models brought forth to explain the Anthropic Principle.

These cosmological models can be divided into two general classes: those which speak of a single universe, in which of course we live; and those which speak of many universes, each of which arises from different initial conditions which determine the values of the constants of nature and the operative physical laws. In this latter case, it is generally supposed that all of those universes which are not self-contradictory (in which the various combinations of constants and physical laws do not defy the principle of contradiction) have actually been realized.

The single-universe cosmologies are several, all of them based on an initial Big Bang, which, in various forms, is up to the present the best explanation of existing observational data. One such cosmological model is that of Stephen Hawking in his book: *A Brief History of Time, From the Big Bang to Black Holes*.[10] From quantum gravity considerations Hawking comes to the conclusion that space-time forms a closed but unbounded surface, and that as such it requires no initial boundary conditions.[11] Hawking says in effect that the only boundary condition is that there are no boundary conditions. Thus the fine-tuned combination of constants of nature and physical laws which eventually led cosmic evolution to the emergence of human civilization is due to nothing other than the inevitable consequences of quantum gravity. Thus, according to Hawking, it is not at all necessary to consider a God-Creator. God is not needed to explain the universe; he does not exist. Leaving aside the purely scientific evaluation of Hawking's theory (very much contested among cosmologists), it is important that his conclusion be evaluated in terms of the principal argument of this paper, namely the confrontation of science and theology, or more specifically, the dialogue between cosmology and theology arising from considerations of the Anthropic Principle. To deny the existence of the God of religious belief on the basis of a scientific theory is a lamentable confusion of two independent ways of knowing. The God of religious belief is not an initial condition, nor even the initial condition, for the existence of the universe. Should, therefore, such a scientific theory really establish that initial conditions are not required, there would still be grounds for science to either affirm or deny the existence of God.

All of the other models of a single universe require the determination of initial conditions from which a certain combination of constants of nature and physical laws came to be so that human civilization evolved. In all of them it is difficult to escape the notion of an arbitrary choice on the part of God, the Creator. It is required, for instance, of God that he has chosen a multitude of precise values for physical constants in such a way that, had he chosen a slightly different value for one constant or other, the evolution of human civilization would not have been possible. God would be, to put it in more pedestrian and vivid terms, somewhat like a master cook whose pinches of salt, sugar, paprika and other ingredients are just right so as to produce the pudding, human civilization. It appears to me that this inevitable inclination to a certain arbitrariness in the religious concept of God-Creator could be removed only if the appropriate cosmological model had built into it all that was necessary to explain scientifically the actual combination of physical laws and constants of nature that we observe. God would, in such a model, not be needed to select the ingredients. This is apparently what Hawking attempts to accomplish in his model derived from quantum gravity considerations. The religious thinker might, of course, be tempted to see this as a threat to the very existence of God, or at least as the establishment of a solipsistic God, completely divorced from the universe. This could only be the case if one seeks to find God through science or seeks to understand the universe through religious thought alone. In either case, as we have noted above in criticizing Hawking, there is crass confusion of epistemologies. On the other hand, if one respects the independent espistemological methodologies of science and theology, but seeks nonetheless for a unity in the human understanding of all reality, then it appears to me that the understanding of God's freedom in the context of single-universe cosmologies is more compatible with the type of model proposed by Hawking.

In considering the many-universe cosmologies it appears that one might arrive at an even more profound compatibility between the religious concept of God-Creator and scientific theories of the origins of the universe, in the sense that God would not be seen either as an arbitrary creator or a solipsist with respect to creation. There are two classes of many-universes: those in which the universes exist simultaneously and those in which they exist successively. For the purposes of this paper I wish to describe briefly one type of each of these two classes. The many-universes could have been born from an initial chaotic state from which there was such a rapid inflationary expansion that the various parts are not able to communicate with one another. Since they cannot communicate among themselves, they can justly be called separate universes. Since they have their respective origins in an initial chaotic state, they each have a different combination of physical laws and constants of nature. Since there are many, if not an infinity of such universes, it is statistically understandable that human civilization exists in one or more, but not in all, of them. Thus the explanation of the Anthropic Principle, even the strong version, requires no intrinsic finality, but has a natural explanation in statistical considerations. I should remark that many scientists consider this version of many-universes, since by definition the many-universes are not verifiable (they are not inter-communicative), to be a non-scientific theory since it does not respect the strict rigours of the scientific method. Another way of obtaining many-universes is by a succession of Big Bangs, that is a series of pulsations in which each cycle of expansion and contraction represents a separate universe. According to the initial conditions at the beginning of each expansion, diverse combinations of physical laws and constants of nature are realized. Again the Anthropic Principle is explained by statistical considerations.

Although further reflection on this issue is surely required, it appears to me initially that the many-universe cosmologies, as compared to the single-universe cosmologies, are more compatible with a God-Creator who is neither arbitrary nor solipsistic. The religious concept would be, for instance, of a God who would have seen his image and likeness emerge in one or more of the many-universes and he would have marvelled, loved, and taken special care of it as he told us he did in his self-revelation in Scripture and Tradition.

7. Further Considerations: Cosmology and Theology

I conclude here, obviously only at the beginning of what could be a rich encounter of theological thought with cosmology. I would like to list, but only as an example, further considerations which might be of interest to those who are much more competent in these areas than I am:

1. How can we express in more detail the concept of God-Creator in terms of either the single-universe or many-universe cosmologies? We have used such words as *see, marvel at, love, have special care for, etc.* Although we wish to avoid having a God who is either arbitrary or solipsist, it does not appear that the above considerations are adequate to express the rich concept of *to create*.
2. How can we preserve the doctrine of a special intervention of God in the creation of the human being[12] without compromising the notion of a free but not arbitrary God in relation to many-universes?
3. If human civilizations exist elsewhere in our universe, or even in other universes, would the Son of God be also incarnate there? Would he have become incarnate among us even if there had not been an original sin?[13] If there were more than one incarnation, how are we to preserve the defined trinitarian and christologican doctrines?
4. In the case of a single universe, how are we to understand the relatively limited time span of a human civilization determined by the ageing of the central star of a planetary system?

In conclusion, I believe that it is quite clear from such considerations as the above that the Anthropic Principle has not only been a stimulus to research in cosmology but that it also provides an exciting point of encounter between theology and the sciences and has surely served to reintegrate the factor human being, which for centuries was excluded from the physical sciences.

References

1. For an analysis of the *Letter to the Duchess Christina* see: J. Dietz Moss, 'The Rhetoric of Proof in Galileo's Writings on the Copernican System', in *The Galileo Affair: A Meeting of Faith and Science*, eds. G.V. Coyne, S.J., M. Heller and J. Zycinski, Specola Vaticana, Città del Vaticano 1985, p. 41.
2. C.P. Snow, *The Two Cultures and the Scientific Revolution*, Cambridge University Press, Cambridge 1959.
3. B. Carter, in *Confrontation of Cosmological Theories with Observation*, ed. M.S. Longair, Reidel, Dordrecht 1974, p. 291.
4. For a recent and detailed discussion of the fine-tuning of the physical constants see: J. Leslie, 'The Prerequisites for Life in the Universe', in *Newton and the New Direction in Science*, eds. G.V. Coyne, S.J., M. Heller and J. Zycinski, Specola Vaticana, Città del Vaticano 1988; also J. Leslie, 'How to Draw Conclusions from a Fine-tuned Universe', in *Physics, Philosophy, and Theology: A Common Quest for Understanding*, eds. R.J. Russell, W.R. Stoeger, S.J., and G.V. Coyne, S.J., Specola Vaticana, Città del Vaticano 1988 and University of Notre Dame Press, Notre Dame 1988, p. 297.
5. See J.D. Barrow and F.J. Tipler, *The Anthropic Principle*, Oxford University Press, Oxford 1986.
6. For a discussion of the classical application of the negative way see note 7. For a review of the use of the negative way in modern theological methods see: A. Beni, 'La Nuova Metodologia Teologica', in *Correnti Teologiche Postconciliari*, ed. A. Marranzini, Città Nuova Editrice, Rome 1974, p. 79.
7. Analogy is here understood in the scholastic sense of the term; see J. Ramirez, 'De analogia secundum doctrinam aristotelico-thomisticam', in *La Ciencia Tomista*, Vol. 244, 20, 1921.
8. In any discussion of the concept of God-Creator one must take into account the rich Scriptural and Patristic tradition of the human being as the image of God. See: F. Festorazzi, 'L'uomo immagine di Dio (Gn 1, 26–17) nel contesto totale della Bibbia', in *Bibbia e Oriente*, Vol. 6, 105, 1964.
9. For an account of recent research on creation in Genesis see: R.J. Clifford, S.J., 'Creation in the Hebrew Bible', *Physics, Philosophy, and Theology, op. cit.*, p. 151.
10. S. Hawking, *A Brief History of Time, From the Big Bang to Black Holes*, Bantam 1988.
11. For a discussion of the concept of initial conditions see: W.R. Stoeger, S.J., 'What Does Science Say About Creation?', *The Month*, 249, 805, 1988.
12. See H. Denziger, *Enchiridion Symbolorum et Definitionum*, 26a, Freiburg 1947: No. 248 for the doctrine of creation in general; Nos. 20 to 24 for the doctrine of the origin of the human soul; Nos. 1910 to 1914 for the doctrine of the union of body and soul.

13. For a recent discussion of the historical debate over the redemptive and purely incarnational aspects of the Incarnation see: J. Galot, 'L'Attuale Problema Cristologico', in *Correnti Teologiche Postconciliari, op. cit.*, p. 204.

Anthropic Arguments – Are They Really Explanations?

BERNULF KANITSCHEIDER

Zentrum für Philosophie und, Grundlagen der Wissenschaft
Justus-Liebig-Universität Gießen

1. The Historical Background

The anthropic principle cannot be understood without its conceptual counterpart, the Copernican principle. Therefore, a short survey of the conceptual development of the problem of the cosmic centre may not be out of place. The displacement of the cosmic centre got along step by step. In 1543, Copernicus shifted the cosmic centre to the sun. At the end of the eighteenth century, Sir William Herschel moved it to the core of our galaxy, the Milky Way. That position was overthrown by Harlow Shapley in 1922, and in 1952 Walter Baade discovered that our galaxy is a typical spiral galaxy without any outstanding characteristics.[1] The upshot of this development can be stated easily: mankind does not occupy the centre of the universe. Its habitation seems to be more or less a typical location in the vastness of space.

Modern cosmology in the time of Einstein and Hubble brought about the idea of the isotropic and homogenous universe. Isotropy and homogeneity act as constraints of Einstein's field equations. They are handled as boundary conditions to get a subclass of world models with uniformly distributed matter content and constant spatial curvature.[2] By observation alone, however, homogeneity is not easily to be gained, only local isotropy can be established by experience. To arrive at global isotropy which in turn entails homogeneity, we need the *Copernican principle*, a recent version of which is called the *location principle*. A concise formulation of this principle has been given by Edward Harrison, when he claimed: 'Distinguished locations are unlikely.'[3]

It can be maintained that modern cosmology works under the paradigm of this principle. In recent times, however, some counter movements against Copernican democracy have been established. The origins of new anthropic ideas are deeply rooted in nineteenth century physics. Ludwig Boltzmann, the founder of statistical mechanics, was led to explain the direction of time by a natural line of argument which comes close to the weak anthropic principle (WAP).[4] His attempt to infer the thermodynamical arrow of time from mechanics forced him to assume one of two possibilities. Either the entire universe is at present in a highly improbable state, or our observable region is a tiny part of the whole universe which globally is in thermodynamic equilibrium. On account of the unavoidable fluctuations, relatively small regions will deviate from thermodynamic equilibrium. The reason why a living being finds himself in such a corner of the universe, for which a significant deviation from equilibrium defines an arrow of time, can be understood if we remember that only on a slope of the entropy curve can life, consciousness and intelligence evolve. Therefore, it does not come as a surprise that in our local cosmical environment time has the direction we actually observe, because open systems need thermodynamic unequilibrium as a necessary condition for the origin, evolution and maintenance of their life.

For the moment, it is worthwhile to notice that Boltzmann had to make the choice, either to accept very improbable initial conditions for which in principle no further experience could be given, or to include man and his thermodynamical presuppositions in a physical argument. Even at that time, Boltzmann's colleagues were reluctant to accept the anthropic fluctuation interpretation and in modern times it was severely criticized by Karl Popper, who accused Boltzmann of having violated the objectivity of the direction of time.[5]

I would like to argue, however, that Boltzmann's intention was in accord with the objectivity of science. He asked whether our habitation is a special part of the total universe, that is constrained by certain conditions of a thermodynamic kind. Even if the universe as a whole is in equilibrium, that part we are living in must be a fluctuation far away from equilibrium because that is a necessary condition for life. It is epistemologically of the uttermost importance to notice that looking for a selection effect is a way of objectifying knowledge, not of subjectifying it. Therefore, Boltzmann used the weak anthropic principle quite in accordance with his overall principles of knowledge which were realistic and objectivistic. The latter can be verified easily from his attacks on the phenomenalistic philosophy of his Vienna colleague Ernst Mach. This historical example is very instructive. It shows that at least the weak anthropic principle can be divorced from teleological, idealistic, and subjectivistic interpretations. I therefore greatly appreciate Brandon Carter's recent proposal to use the term 'congnizability principle' instead of 'anthropic principle'.[6]

A second point should be noticed. The Copernican and the anthropic principle have been construed mostly in a very antagonistic way. Cosmology may convince us that from a global view-point our kind of intelligent life has no special *location* in the universe. But on a local perspective, we are surrounded by characteristic types of objects that might be unique. Our planetary system, a special location, where the only known biological evolution took place, is perhaps distinguished in so far as peculiar constraints led to the building site of the biosphere. Current theories on the origin of life give no hint on the intrinsic properties of possible organisms that might be engendered by other planetary sites. The reason is, beside the low state of theoretical knowledge on the origin of life, that cosmology gives us only the main outline of the cosmic picture and ignores the irregularities of the local detail. Even a globally homogeneous universe without cosmic centre and without outer edge can have a physically privileged place where evolution took place. Cosmology makes spatio-temporal assertions, it is not concerned with estimations of complexity and *a fortiori* not with statements of value.[7] Accordingly, it is possible that the human neural network is the system with the highest complexity in the entire universe and the only system that engenders values. This is fully compatible with our living in a typical location of the universe.

2. Anthropocentricity, Teleology, and Evolution

In modern science, teleological explanations had to give way to causal mechanistic explanations almost everywhere. Darwin's evolutionary theory expelled the question from the domain of science, whether developmental trends of organic systems are goal-directed. *Global teleology*, the long-range purposeful development of nature, has been revealed as a myth. *Local teleological behaviour*, e.g. planned action or free voluntary decision, has been restricted to a rather small subclass of higher mammals. Goal-directed actions of higher animals are, of course, not at variance with a causal description of these processes as certain activations of the central nervous system. They can be reconstructed properly as emergent qualities of a higher level of complex organizations.

Darwin's mechanistic, that means dysteleological, model of explanation has been worked out in two directions. Even the topic of the origin of life came under the domain of evolutionary thinking. A hundred years later, in 1971, Manfred Eigen showed a possible way to the 'self-organization of matter and the evolution of biological macromolecules'.[8] Herewith the outline of a theory describing the physical origin of life as getting out from a molecular chaos was at hand. On the other side, Edward O. Wilson, almost at the same time, extended the Darwinian model up to the level of human social behaviour and even up to ethics.[9] The naturalistic approach to the mental functions of man has gained strength with the enormous advances in the brain sciences. A novel idea got more and more shape and directness: every human achievement including mind and culture has a material base and originated during the evolution of the human genetic constitution and its interaction with the environment. Edward Wilson and Michael Ruse have put it succinctly: 'We suggest that it will prove possible to proceed from a knowledge of the material basis of moral feeling to general accepted rules of conduct.'[10]

Within this trend to a naturalistic world picture, the currently much debated evolutionary epistemology is only the last rung of a ladder leading up from the domain of physics to conscious and intelligent systems with their cognitive degrees of freedom.[11]

Given this unmistakable historical trend of natural science, it is more than surprising that even within pure physics (e.g. cosmology and astrophysics) apparent teleological explanations came to the fore. Some physicists like Barry Collins and Stephen Hawking have been rather silent on the correct interpretation of the use of their anthropic arguments. The last sentence of their seminal paper on the isotropy of the universe[12] states '… the answer to the question, why is the universe isotropic is, because we are here'. Other authors like Robert Dicke and P.J.E. Peebles are more explicit on the logics of the relationship between man and the universe within an anthropic argument. They ask: 'Could it be that it is the presence of observers that determines the nature of the universe?'[13] Now the crucial question is, how to interpret the term 'determines'. Does it really point to a goal-directed process? The case of Collins' and Hawking's argument seems to suggest nothing like that. It is, of course, a surprising fact that isotropy correlates strongly with the existence of intelligent life. That astonishment can be diminished when the additional hypothesis is taken into account, that reality consists not only of one universe but of an infinite ensemble of universes with all possible initial conditions. Nearly each of these worlds becomes highly anisotropic at large times and therefore contains no observer. In this case no Aristotelian 'entelechia' is needed to understand the amazing coincidence. It just mirrors the selection effect of our own existence. It is part and parcel of the customary scientific methodology to be aware of selection effects that can bias our observation. Taking cognizance of this selection effect which carbonaceous astronomers force upon the set of physically possible worlds may indeed reduce the surprise on the *a priori* improbably features of nature. But, as I will argue in due time, taking account of this selection effect cannot substitute for causal explanations, why these properties of the universe prevail and just yield the necessary conditions for carbon-based life.

3. Are Anthropic Arguments Explanatory?

As I understand the anthropic principles, their right and legitimate use is to regard them as a blank to be filled in by causal connections of the pertinent facts.[14] It is of course most difficult to establish causal interpretations, if the so-called strong anthropic principle (SAP) is taken into account, which, in 1974 Brandon Carter defined in the following way: 'The universe *must be* such as to admit the creation of observers within it at some stage.' Here the term 'must' is of crucial importance and it is of course open to a variety of meanings. If we take it to imply coercion, then the latter points to an entity or agent that compels or forces the constants of nature to take their life-sustaining values. In this case, the teleological interpretation seems to be inescapable and this is the meaning many have made out of the strong anthropic principle. E.g., we can read in a book of Paul Davies the following reconstruction of the SAP: 'In essence, the strong anthropic principle claims that the universe is tailor-made for habitation and that both the laws of physics and the initial conditions obligingly arrange themselves in such a way that living organisms are subsequently assured of existence.'[15] Of course, it is not a logical consequence in the sense of the entailment relation to take the strong anthropic principle in this way. But it is difficult to see how it can be interpreted otherwise. Within the framework of naturalistic science it is difficult to make sense out of the above-mentioned claim. Why should a material universe be tailor-made for habitation?; what does it mean that the initial conditions are obliged to arrange themselves in a goal-directed way which points to the evolution of man? These moral terms are undefined before the advent of intelligent beings. They are predications devoid of physical meaning within the context of naturalistic ontology. Barrow and Tipler have rightly remarked that the design version of the strong anthropic principle can only be understood if couched in a theological language that transcends the limits of science. In the same way, the so-called participatory version of the strong anthropic principle cannot be implemented into current naturalistic science. The participatory anthropic principle (PAP) says, in the definition of Barrow and Tipler: 'Observers are necessary to bring the universe into being.'[16] The PAP is motivated clearly by quantum mechanics and by an extreme epistemological idealism.[17] Taken at the face value, its claim is stronger than the customary phenomenalism of the Machian type. Phenomenalism asserts that the only object of our investigations can be the world of our sense experiences, but not the way they are brought about by the physical objects behind the phenomena. A causal theory of sense experience is accordingly impossible. Taken verbally, the participatory anthropic principle states even more than phenomenalism, namely the causal efficacy of observers in the universe. Observers seem to engender or produce physical reality. Needless to say, we do not have the slightest hint of such a kind of retrocausal activity of intelligent organisms. Cognition as it is understood today is an activity of the neural network and it is in no way exempted from customary time asymmetric causality. No neurological model could be established within current scientific theories to fill in the mechanism – responsible for a process of cognition – which, at cosmologically late times, arranges the constants and parameters of the early universe. That conjecture has trespassed the borderline between science and speculative metaphysics.[18]

Given the four main varieties of the anthropic principle, it is of central importance to realize which of them can be defended considering the approved rules of today's scientific methodology. Philosophers of science have uttered heavy criticisms on the various anthropic principles, foremost when they are put forward as explanations of a new anthropocentric type. The patterns of explanations are thought to be well understood, since C.G. Hempel analysed them as the deduction of an event E (explanandum) from a set of law statements together with some initial and boundary conditions.[19] If we take the argument of Collins and Hawking as our main example, then the decisive boundary condition is that 'intelligent life exists'. Given the law of Einstein's gravitation theory, some additional astrophysical assumptions and some biological generalizations, we can construct a formally correct explanation of the isotropy of the universe. Nevertheless, its direction of explanation is wrong. The causal connection is *one way*, namely from the isotropy of the universe at earlier times to the existence of intelligent organisms at later times, but not the other way round.[20] The violation of such a global property of spacetime, as the causal structure, makes the anthropic explanation an illegitimate use of the anthropic principle. The methodological moral of this example is well-known: Logic alone is not enough to identify correct explanations. An explanatory argument has not only to be deductively correct, it needs to be supplemented by synthetic rules of a more general kind, pertaining to our factual world.[21]

Anthropic arguments can, of course, point to novel, hitherto unknown concatenations of organic systems with their cosmological embedding. If, however, answers to questions of the type

Why is it that 3-space is flat, or
3 K radiation is isotropic, or
the universe is as old as a main sequence star (10^{10} y), or
the universe is roughly 10^{10} light years in extent?

are requested, then we have to take care of the direction of explanation. We are of course allowed to add further premises like the postulate of the world ensemble. If the enlargement of physical ontology has been made acceptable on philosophical reasons,[22] an answer as to why we observe something that is of infinitesimal probability can be given without inverting the causal structure of physical processes. In principle, a causal analysis should be feasible for every member of the world ensemble. Therefore, I do not like to adopt the stance of D. Lewis, who argues that the postulate of the world ensemble is a reason why we do not need an explanation at all. I concede that the ensemble hypothesis reduces the astonishment on the improbability, but I hold that it cannot be taken as a substitute for the causal explanation of why, in the individual worlds, the constants arrange themselves in the way they do.

Some philosophers have taken the point of view that even extremely rare coincidences do not need an explanation at all, so why bother with improbable chance events? Since unlikely events occur sometimes, we should not be disturbed by the above-mentioned cosmological coincidences. John Leslie has rightly criticized this kind of shrugging one's shoulders.[23] Take an example: If, on a lonely beach, you come across an inscription several square feet in extent 'Coca Cola', you would not argue that the written character 'Coca Cola' is one of the many possible combinations of sand grains occurring in the course of long geological times. Given the existence of humans, everybody would tend to a causal explanation: A stroller on the beach must have written these two words, maybe as a joke. If, however, no organism is available for a dynamical explanation, a natural process has to be found to make this contrived order of sand grains comprehensible. I will give a sketch of this solution in the last section. John Leslie's own philosophical approach which he calls 'extreme axiarchism' and which contains a 'creative efficacy of ethical requiredness'[24] appears to me of dubious ontological status. It seems far more difficult to vindicate philosophically a Platonic existence of values than the existence of many, even infinitely many, physical worlds.

4. Anthropic and Dynamic Explanations

From an epistemological point of view the central question seems to be how to get out of this limbo of metaphysical ideas. I hold that, for epistemological reasons, neither anthropocentric world views nor Platonic values in themselves nor teleological supernaturalism can be accepted as explanatory strategies within physics. They may be held as personal knowledge, but they cannot be objectified and therefore cannot be included in a scientific approach to nature. Therefore, I put my money elsewhere. My hope for an explanation of the cosmic coincidences constraining the existence of intelligent life derives from the development of the Grand Unified Theories (GUT). This approach has the great advantage that it operates within the well-corroborated paradigm of science, without taking refuge in exotic modes of explanation.[25] This is a conservative stand, but it leads to a causal understanding of the *a priori* improbable combination of parameters and constants within the received view of science. The GUTs are designed to solve the deepest questions of today's physics concerning the problem of contingency. Physics should deliver good reasons as to why the constants of nature have their special values. We would like to understand the coupling constants of the different forces, the enigma of the vanishing cosmological constant λ (in natural units $|\lambda| = 10^{-120}$), the distinguished value of the density parameter $\Omega_0 = 1$, which makes anisotropy distortions damp away at very late times and gives us the unique property of isotropy. These are contingent features of our world to be explained by a strong theory. Above that, physics should give us a reason, why spacetime is four dimensional, space has three dimensions and time only one dimension. These questions cannot be tackled within the cosmological standard model, but a few of them – although not the λ-problem – can be solved within the unified theory approach. E.g., the inflationary cosmology can answer the question why certain crucial life-sustaining features are realized in our universe. As a common trait, all inflationary models assume the existence of some stage of evolution at which the universe expands exponentially. The exponential expansion dominating the universe for a short period characterized by the deSitter line element is caused by the vacuum energy. The decay of the vacuum is the crucial ingredient of the inflationary paradigm. The value $\Omega = 1$ is a strong constraint for the origin and the evolution of intelligent life. In the frame of the inflationary scenario, the present proximity of the universe to the critical density becomes comprehensible by a causal process. The huge exponential expansion caused the curvature to become negligible, as we know from the surface of a rubber balloon which is heavily inflated.[26]

There is also a possible answer to a special case of the contingency problem forming a main part in the discussion on the anthropic principle, namely the dimensionality of spacetime. Already before the advent of the anthropic principles, G.J. Whitrow tried to elucidate the three dimensionality of space in an anthropic manner. He argued by means of a counterfactual analysis: If the dimension of space n would be $n > 3$, there could not exist stable orbits for planetary sites which are possible habitations for humans. The dimensions $n = 1$ and $n = 2$ are excluded, because such spaces do not allow the concatenation of a large number of nerve cells in order to form a complex neural network that alone, to our knowledge, is able to produce mind generating ideas. There is no doubt that Whitrow uses the argument in an explanatory way, the existence of a 'minding animal' among other premises acting the part of an explanans for the three dimensionality of space. Says Whitrow: '... that the number of dimensions of space is necessarily three, no more and no less, because it is the unique natural concomitant of the higher form of terrestrial life'.[27] Rightly, J.J.C. Smart has criticized the back to front explanatory use of an anthropic argument as being preposterous.[28] On this point, it is my conviction that a stronger theory has to fill the bill. Some recent unified theories contain a renewal of an old idea of Theodor Kaluza and Oskar Klein. They introduced theories with spacetime of more than four dimensions. These extra dimensions of space – for instance, superstring theories are favourably defined in ten dimensions – have been compactified, that means, they have shrunk into thin tubes of the order of the Planck length. According to such theories, the universe had its full dimensionality at the unification energy of $T = 10^{17}$ GeV.

Today all dimensions except our well-known three spatial dimensions have been confined to experimentally unfathomable small regions. Here a road opens to a much deeper understanding of the old question, as to why space has exactly three and spacetime, accordingly, four dimensions. It is *because* – and here we can use this term in the normal explanatory sense – today our universe is cold so that the higher dimensions of space have been frozen up.

A possible critique of this approach might argue that the problem has only been deferred to the adjacent question, why the ten dimensions of spacetime existed at that early time. But this point seems to me irrelevant. Every explanation has to start somewhere, nothing can be inferred from the zero set of presuppositions. Therefore, the compactification process is a step to a deeper understanding, anyway. Moreover, there are good mathematical reasons stemming from the symmetry group of the theories, which give a rational defence as to why the superstring theory is properly defined in ten dimensions. On the other side, there are genuine questions on the Kaluza/Klein approach which cannot be put aside so easily. We may wonder why compactification stopped exactly with four dimensions of spacetime and not with any other number. Furthermore, there should be a proof of uniqueness of a certain compactification scheme giving us the distinguished road from the full dimensionality of the high energy range down to our four dimensions of low energy particle physics. Most desirable, of course, would be a theory in which only one type of compactification can lead inevitably to our four dimensional inflationary universe together with the low energy physics we encounter in the present experiments and observations.

To advocate dynamical explanations should not be taken as to devalue the anthropic interconnections. The innumerable links that have been found between intelligent life and astrophysical and cosmological facts remind us how strongly we are bound to our large-scale environment. Set in its proper stage the anthropic principle is to be seen in the context of the unity of the universe. It seems plausible to me to see the anthropic principle in an older tradition of ideas, namely the metaphysics of Leibniz' monadology, which says roughly that every element of the universe contains in itself a mirror-like picture of the cosmic ground-plan. Since the anthropic principle connects the properties of organisms to certain traits of the cosmic environment, it can be called a partial realization of the bootstrap idea. In the words of Edward Harrison, 'the anthropic principle serves as a make-shift' and can be called 'the poor man's bootstrap'.[29]

On the other side, the anthropic principle is not an invitation to turn the customary direction of explanation upside down, but to fill in the blank of dynamical considerations. Dynamical explanations by unified theories lead into deep philosophical waters. They contain surplus meaning in so far as they not only allow the deduction of the anthropic constraints and numerical coincidences, but beyond that convey answers to questions physicists would never have imagined to bring into the domain of rational science.

References

1. Cf. to this problem the now classical analysis of D.W. Sciama: *The Unity of the Universe*, London 1959.
2. After their discoverers, these models are called Freidmann–Robertson–Walker spacetimes or FRW worlds for short.
3. E. Harrison: *Cosmology*, Cambridge 1981, p. 90.
4. L. Boltzmann: *Lectures on Gas Theory*, transl. by St. G. Brush, Berkeley 1964, p. 446.
5. K.R. Popper: 'Autobiographical Notes', in P.A. Schilpp (ed.): *The Philosophy of Karl Popper*, LaSalle (Ill.), p. 126.
6. B. Carter: 'The Anthropic Principle and its Implications for Biological Evolution', *Phil. Trans. Roy. Soc. Lond.* A **310**, p. 347.
7. Accordingly, it makes no sense to talk of something like 'the meaning of the universe', because it is impossible to define 'meaning' on the physical level of description.
8. M. Eigen: 'Selforganization of Matter and the Evolution of Biological Macromolecules', *Die Naturwissenschaften* **58** (1971), p. 465.
9. E.O. Wilson: *On Human Nature*, Cambridge (Mass.) 1978.
10. E.O. Wilson and M. Ruse: *Biology and Philosophy*.
11. G. Radnitzky and W.W. Bartley III: *Evolutionary Epistemology, Rationality, and the Sociology of Knowledge*, Open Court LaSalle, Illinois 1987.
12. B. Collins and Stephen Hawking: 'Why is the Universe Isotropic?', *Astrophys. Journ.* **180** (1973), pp. 317–34.
13. R. Dicke and P.J.E. Peebles: 'The Big Bang Cosmology – Enigmas and Nostrums', in: S.W. Hawking and P.J.E. Peebles: *General Relativity*, Cambridge 1979, pp. 504–17.
14. This demand should be taken as a methodological ideal, as a rule and not as a dogmatic principle. It can only be fulfilled if no pure random processes are involved. It is self-evident that we cannot force upon nature law formulae that are more strongly deterministic than the inner structure of nature allows. In that case the anthropic understanding of the coincidence would be *ultima ratio* that cannot be trespassed.
15. P.C.W. Davies: *The Accidental Universe*, Cambridge University Press 1987, p. 120.
16. J.D. Barrow and F. Tipler: *The Anthropic Cosmological Principle*, Oxford 1986, p. 22.
17. There is no generally accepted interpretation of QM. A variety of conceptual schemes compete to yield the genuine interpretation. Among them, the so-called orthodox subjectivistic version has been used in some quarters for a support of an idealistic physical world view. The PAP is without doubt an outcome of this idealistic perspective.
18. You can of course question even the naturalistic basis of modern science, but

since it is an outcome of a few hundred years of scientific enquiry into the building blocks of nature we should be very cautious when taking into consideration such a revolution in scientific ontology.

19. C.G. Hempel: *Aspects of Scientific Explanation and other Essays*, New York 1964.

20. This point was made first by J.J.C. Smart: 'Philosophical Problems of Cosmology', *Revue International de Philosophie* **160** (1987), pp. 112–26.

21. There are plenty of examples that show how insufficient logic is to give a complete account of a scientific explanation. Consider e.g. the example of Sylvain Bromberger: The height of a flag pole together with the angle of the sunlight falling on it explains the length of the shadow on the ground, but not vice versa! Of course, we cannot explain the height of the flag pole by the length of its shadow on the ground and a certain angle of the sunlight.

22. Questions may arise as to how large this ensemble will be. Does it contain all logical possible universes, or only those physically possible worlds which are allowed by present natural laws? If the first alternative is requested, we may run into the risk of a runaway ontology, insofar as the power set of the ensemble should exist and the class of existing objects is growing without limit.

23. J. Leslie: 'Anthropic Principles, World Ensemble, Design', *Am. Phil. Quat.* **19** (1982), pp. 141–51.

24. J. Leslie: *Value and Existence*, Oxford 1979.

25. Compare: B. Kanitscheider: 'Explanation in Physical Cosmology', *Etkenntnis* **22** (1985), pp. 253–63.

26. This example, however, contains the empirical problem, that present baryon density is far below the critical density by a factor of 10. If the dark matter cannot be found, be it as massive neutrinos or as exotic particles like axions, this argument is less cogent.

27. G.J. Whitrow: 'Why Physical Space has Three Dimensions', *Brit. Journ. Phil. Sci.* **6** (1955), pp. 13–31.

28. J.J.C. Smart: a.a.O., p. 113.

29. E. Harrison: *Cosmology*, Cambridge 1981, p. 115.